Fractals: A Very Short Introduction

VERY SHORT INTRODUCTIONS are for anyone wanting a stimulating and accessible way into a new subject. They are written by experts, and have been translated into more than 45 different languages.

The series began in 1995, and now covers a wide variety of topics in every discipline. The VSI library now contains over 500 volumes—a Very Short Introduction to everything from Psychology and Philosophy of Science to American History and Relativity—and continues to grow in every subject area.

Titles in the series include the following:

AFRICAN HISTORY John Parker and Richard Rathbone
AGEING Nancy A. Pachana
AGNOSTICISM Robin Le Poidevin
AGRICULTURE Paul Brassley and Richard Soffe
ALEXANDER THE GREAT Hugh Bowden
ALGEBRA Peter M. Higgins
AMERICAN HISTORY Paul S. Boyer
AMERICAN IMMIGRATION David A. Gerber
AMERICAN LEGAL HISTORY G. Edward White
AMERICAN POLITICAL HISTORY Donald Critchlow
AMERICAN POLITICAL PARTIES AND ELECTIONS L. Sandy Maisel
AMERICAN POLITICS Richard M. Valelly
THE AMERICAN PRESIDENCY Charles O. Jones
AMERICAN SLAVERY Heather Andrea Williams
THE AMERICAN WEST Stephen Aron
AMERICAN WOMEN'S HISTORY Susan Ware
ANAESTHESIA Aidan O'Donnell
ANARCHISM Colin Ward
ANCIENT EGYPT Ian Shaw
ANCIENT GREECE Paul Cartledge
THE ANCIENT NEAR EAST Amanda H. Podany
ANCIENT PHILOSOPHY Julia Annas
ANCIENT WARFARE Harry Sidebottom
ANGLICANISM Mark Chapman
THE ANGLO-SAXON AGE John Blair
ANIMAL BEHAVIOUR Tristram D. Wyatt
ANIMAL RIGHTS David DeGrazia
ANXIETY Daniel Freeman and Jason Freeman
ARCHAEOLOGY Paul Bahn
ARISTOTLE Jonathan Barnes
ART HISTORY Dana Arnold
ART THEORY Cynthia Freeland
ASTROPHYSICS James Binney
ATHEISM Julian Baggini
THE ATMOSPHERE Paul I. Palmer
AUGUSTINE Henry Chadwick
THE AZTECS David Carrasco
BABYLONIA Trevor Bryce
BACTERIA Sebastian G. B. Amyes
BANKING John Goddard and John O. S. Wilson
BARTHES Jonathan Culler
BEAUTY Roger Scruton
THE BIBLE John Riches
BLACK HOLES Katherine Blundell
BLOOD Chris Cooper
THE BODY Chris Shilling
THE BOOK OF MORMON Terryl Givens
BORDERS Alexander C. Diener and Joshua Hagen
THE BRAIN Michael O'Shea
THE BRICS Andrew F. Cooper
BRITISH POLITICS Anthony Wright

BUDDHA Michael Carrithers
BUDDHISM Damien Keown
BUDDHIST ETHICS Damien Keown
BYZANTIUM Peter Sarris
CANCER Nicholas James
CAPITALISM James Fulcher
CATHOLICISM Gerald O'Collins
CAUSATION Stephen Mumford and Rani Lill Anjum
THE CELL Terence Allen and Graham Cowling
THE CELTS Barry Cunliffe
CHEMISTRY Peter Atkins
CHILD PSYCHOLOGY Usha Goswami
CHINESE LITERATURE Sabina Knight
CHOICE THEORY Michael Allingham
CHRISTIAN ART Beth Williamson
CHRISTIAN ETHICS D. Stephen Long
CHRISTIANITY Linda Woodhead
CIRCADIAN RHYTHMS Russell Foster and Leon Kreitzman
CITIZENSHIP Richard Bellamy
CLASSICAL MYTHOLOGY Helen Morales
CLASSICS Mary Beard and John Henderson
CLIMATE Mark Maslin
CLIMATE CHANGE Mark Maslin
COGNITIVE NEUROSCIENCE Richard Passingham
THE COLD WAR Robert McMahon
COLONIAL AMERICA Alan Taylor
COLONIAL LATIN AMERICAN LITERATURE Rolena Adorno
COMBINATORICS Robin Wilson
COMMUNISM Leslie Holmes
COMPLEXITY John H. Holland
THE COMPUTER Darrel Ince
COMPUTER SCIENCE Subrata Dasgupta
CONFUCIANISM Daniel K. Gardner
CONSCIOUSNESS Susan Blackmore
CONTEMPORARY ART Julian Stallabrass
CONTEMPORARY FICTION Robert Eaglestone
CONTINENTAL PHILOSOPHY Simon Critchley
CORAL REEFS Charles Sheppard
CORPORATE SOCIAL RESPONSIBILITY Jeremy Moon
COSMOLOGY Peter Coles
CRIMINAL JUSTICE Julian V. Roberts
CRITICAL THEORY Stephen Eric Bronner
THE CRUSADES Christopher Tyerman
CRYSTALLOGRAPHY A. M. Glazer
DADA AND SURREALISM David Hopkins
DANTE Peter Hainsworth and David Robey
DARWIN Jonathan Howard
THE DEAD SEA SCROLLS Timothy Lim
DECOLONIZATION Dane Kennedy
DEMOCRACY Bernard Crick
DEPRESSION Jan Scott and Mary Jane Tacchi
DERRIDA Simon Glendinning
DESIGN John Heskett
DEVELOPMENTAL BIOLOGY Lewis Wolpert
DIASPORA Kevin Kenny
DINOSAURS David Norman
DREAMING J. Allan Hobson
DRUGS Les Iversen
DRUIDS Barry Cunliffe
THE EARTH Martin Redfern
EARTH SYSTEM SCIENCE Tim Lenton
ECONOMICS Partha Dasgupta
EGYPTIAN MYTH Geraldine Pinch
EIGHTEENTH-CENTURY BRITAIN Paul Langford
THE ELEMENTS Philip Ball
EMOTION Dylan Evans
EMPIRE Stephen Howe
ENGLISH LITERATURE Jonathan Bate
THE ENLIGHTENMENT John Robertson
ENVIRONMENTAL ECONOMICS Stephen Smith
ENVIRONMENTAL POLITICS Andrew Dobson
EPICUREANISM Catherine Wilson
EPIDEMIOLOGY Rodolfo Saracci
ETHICS Simon Blackburn
THE ETRUSCANS Christopher Smith
EUGENICS Philippa Levine
THE EUROPEAN UNION John Pinder and Simon Usherwood
EVOLUTION Brian and Deborah Charlesworth

EXISTENTIALISM Thomas Flynn
THE EYE Michael Land
FAMILY LAW Jonathan Herring
FASCISM Kevin Passmore
FEMINISM Margaret Walters
FILM Michael Wood
FILM MUSIC Kathryn Kalinak
THE FIRST WORLD WAR Michael Howard
FOOD John Krebs
FORENSIC PSYCHOLOGY David Canter
FORENSIC SCIENCE Jim Fraser
FORESTS Jaboury Ghazoul
FOSSILS Keith Thomson
FOUCAULT Gary Gutting
FREE SPEECH Nigel Warburton
FREE WILL Thomas Pink
FREUD Anthony Storr
FUNDAMENTALISM Malise Ruthven
FUNGI Nicholas P. Money
THE FUTURE Jennifer M. Gidley
GALAXIES John Gribbin
GALILEO Stillman Drake
GAME THEORY Ken Binmore
GANDHI Bhikhu Parekh
GEOGRAPHY John Matthews and David Herbert
GEOPOLITICS Klaus Dodds
GERMAN LITERATURE Nicholas Boyle
GERMAN PHILOSOPHY Andrew Bowie
GLOBAL CATASTROPHES Bill McGuire
GLOBAL ECONOMIC HISTORY Robert C. Allen
GLOBALIZATION Manfred Steger
GOD John Bowker
GRAVITY Timothy Clifton
THE GREAT DEPRESSION AND THE NEW DEAL Eric Rauchway
HABERMAS James Gordon Finlayson
THE HEBREW BIBLE AS LITERATURE Tod Linafelt
HEGEL Peter Singer
HERODOTUS Jennifer T. Roberts
HIEROGLYPHS Penelope Wilson
HINDUISM Kim Knott
HISTORY John H. Arnold
THE HISTORY OF ASTRONOMY Michael Hoskin
THE HISTORY OF CHEMISTRY William H. Brock
THE HISTORY OF LIFE Michael Benton
THE HISTORY OF MATHEMATICS Jacqueline Stedall
THE HISTORY OF MEDICINE William Bynum
THE HISTORY OF TIME Leofranc Holford-Strevens
HIV AND AIDS Alan Whiteside
HOLLYWOOD Peter Decherney
HUMAN ANATOMY Leslie Klenerman
HUMAN EVOLUTION Bernard Wood
HUMAN RIGHTS Andrew Clapham
THE ICE AGE Jamie Woodward
IDEOLOGY Michael Freeden
INDIAN PHILOSOPHY Sue Hamilton
THE INDUSTRIAL REVOLUTION Robert C. Allen
INFECTIOUS DISEASE Marta L. Wayne and Benjamin M. Bolker
INFINITY Ian Stewart
INFORMATION Luciano Floridi
INNOVATION Mark Dodgson and David Gann
INTELLIGENCE Ian J. Deary
INTERNATIONAL MIGRATION Khalid Koser
INTERNATIONAL RELATIONS Paul Wilkinson
IRAN Ali M. Ansari
ISLAM Malise Ruthven
ISLAMIC HISTORY Adam Silverstein
ISOTOPES Rob Ellam
ITALIAN LITERATURE Peter Hainsworth and David Robey
JESUS Richard Bauckham
JOURNALISM Ian Hargreaves
JUDAISM Norman Solomon
JUNG Anthony Stevens
KABBALAH Joseph Dan
KANT Roger Scruton
KNOWLEDGE Jennifer Nagel
THE KORAN Michael Cook
LATE ANTIQUITY Gillian Clark
LAW Raymond Wacks
THE LAWS OF THERMODYNAMICS Peter Atkins
LEADERSHIP Keith Grint
LEARNING Mark Haselgrove

LEIBNIZ Maria Rosa Antognazza
LIBERALISM Michael Freeden
LIGHT Ian Walmsley
LINGUISTICS Peter Matthews
LITERARY THEORY Jonathan Culler
LOCKE John Dunn
LOGIC Graham Priest
MACHIAVELLI Quentin Skinner
MAGIC Owen Davies
MAGNA CARTA Nicholas Vincent
MAGNETISM Stephen Blundell
MARINE BIOLOGY Philip V. Mladenov
MARTIN LUTHER Scott H. Hendrix
MARTYRDOM Jolyon Mitchell
MARX Peter Singer
MATERIALS Christopher Hall
MATHEMATICS Timothy Gowers
THE MEANING OF LIFE Terry Eagleton
MEASUREMENT David Hand
MEDICAL ETHICS Tony Hope
MEDIEVAL BRITAIN John Gillingham and Ralph A. Griffiths
MEDIEVAL LITERATURE Elaine Treharne
MEDIEVAL PHILOSOPHY John Marenbon
MEMORY Jonathan K. Foster
METAPHYSICS Stephen Mumford
MICROBIOLOGY Nicholas P. Money
MICROECONOMICS Avinash Dixit
MICROSCOPY Terence Allen
THE MIDDLE AGES Miri Rubin
MILITARY JUSTICE Eugene R. Fidell
MINERALS David Vaughan
MODERN ART David Cottington
MODERN CHINA Rana Mitter
MODERN FRANCE Vanessa R. Schwartz
MODERN IRELAND Senia Pašeta
MODERN ITALY Anna Cento Bull
MODERN JAPAN Christopher Goto-Jones
MODERNISM Christopher Butler
MOLECULAR BIOLOGY Aysha Divan and Janice A. Royds
MOLECULES Philip Ball
MOONS David A. Rothery
MOUNTAINS Martin F. Price
MUHAMMAD Jonathan A. C. Brown
MUSIC Nicholas Cook
MYTH Robert A. Segal
THE NAPOLEONIC WARS Mike Rapport
NELSON MANDELA Elleke Boehmer
NEOLIBERALISM Manfred Steger and Ravi Roy
NEWTON Robert Iliffe
NIETZSCHE Michael Tanner
NINETEENTH-CENTURY BRITAIN Christopher Harvie and H. C. G. Matthew
NORTH AMERICAN INDIANS Theda Perdue and Michael D. Green
NORTHERN IRELAND Marc Mulholland
NOTHING Frank Close
NUCLEAR PHYSICS Frank Close
NUMBERS Peter M. Higgins
NUTRITION David A. Bender
THE OLD TESTAMENT Michael D. Coogan
ORGANIC CHEMISTRY Graham Patrick
THE PALESTINIAN-ISRAELI CONFLICT Martin Bunton
PANDEMICS Christian W. McMillen
PARTICLE PHYSICS Frank Close
THE PERIODIC TABLE Eric R. Scerri
PHILOSOPHY Edward Craig
PHILOSOPHY IN THE ISLAMIC WORLD Peter Adamson
PHILOSOPHY OF LAW Raymond Wacks
PHILOSOPHY OF SCIENCE Samir Okasha
PHOTOGRAPHY Steve Edwards
PHYSICAL CHEMISTRY Peter Atkins
PILGRIMAGE Ian Reader
PLAGUE Paul Slack
PLANETS David A. Rothery
PLANTS Timothy Walker
PLATE TECTONICS Peter Molnar
PLATO Julia Annas
POLITICAL PHILOSOPHY David Miller
POLITICS Kenneth Minogue
POPULISM Cas Mudde and Cristóbal Rovira Kaltwasser
POSTCOLONIALISM Robert Young

POSTMODERNISM Christopher Butler
POSTSTRUCTURALISM Catherine Belsey
PREHISTORY Chris Gosden
PRESOCRATIC PHILOSOPHY Catherine Osborne
PRIVACY Raymond Wacks
PSYCHIATRY Tom Burns
PSYCHOLOGY Gillian Butler and Freda McManus
PSYCHOTHERAPY Tom Burns and Eva Burns-Lundgren
PUBLIC ADMINISTRATION Stella Z. Theodoulou and Ravi K. Roy
PUBLIC HEALTH Virginia Berridge
QUANTUM THEORY John Polkinghorne
RACISM Ali Rattansi
REALITY Jan Westerhoff
THE REFORMATION Peter Marshall
RELATIVITY Russell Stannard
RELIGION IN AMERICA Timothy Beal
THE RENAISSANCE Jerry Brotton
RENAISSANCE ART Geraldine A. Johnson
REVOLUTIONS Jack A. Goldstone
RHETORIC Richard Toye
RISK Baruch Fischhoff and John Kadvany
RITUAL Barry Stephenson
RIVERS Nick Middleton
ROBOTICS Alan Winfield
ROMAN BRITAIN Peter Salway
THE ROMAN EMPIRE Christopher Kelly
THE ROMAN REPUBLIC David M. Gwynn
RUSSIAN HISTORY Geoffrey Hosking
RUSSIAN LITERATURE Catriona Kelly
THE RUSSIAN REVOLUTION S. A. Smith
SAVANNAS Peter A. Furley
SCHIZOPHRENIA Chris Frith and Eve Johnstone
SCIENCE AND RELIGION Thomas Dixon
THE SCIENTIFIC REVOLUTION Lawrence M. Principe
SCOTLAND Rab Houston
SEXUALITY Véronique Mottier
SHAKESPEARE'S COMEDIES Bart van Es
SIKHISM Eleanor Nesbitt
SLEEP Steven W. Lockley and Russell G. Foster
SOCIAL AND CULTURAL ANTHROPOLOGY John Monaghan and Peter Just
SOCIAL PSYCHOLOGY Richard J. Crisp
SOCIAL WORK Sally Holland and Jonathan Scourfield
SOCIALISM Michael Newman
SOCIOLOGY Steve Bruce
SOCRATES C. C. W. Taylor
SOUND Mike Goldsmith
THE SOVIET UNION Stephen Lovell
THE SPANISH CIVIL WAR Helen Graham
SPANISH LITERATURE Jo Labanyi
SPORT Mike Cronin
STARS Andrew King
STATISTICS David J. Hand
STUART BRITAIN John Morrill
SYMMETRY Ian Stewart
TAXATION Stephen Smith
TELESCOPES Geoff Cottrell
TERRORISM Charles Townshend
THEOLOGY David F. Ford
TIBETAN BUDDHISM Matthew T. Kapstein
THE TROJAN WAR Eric H. Cline
THE TUDORS John Guy
TWENTIETH-CENTURY BRITAIN Kenneth O. Morgan
THE UNITED NATIONS Jussi M. Hanhimäki
THE U.S. CONGRESS Donald A. Ritchie
THE U.S. SUPREME COURT Linda Greenhouse
THE VIKINGS Julian Richards
VIRUSES Dorothy H. Crawford
VOLTAIRE Nicholas Cronk
WAR AND TECHNOLOGY Alex Roland
WATER John Finney
WILLIAM SHAKESPEARE Stanley Wells
WITCHCRAFT Malcolm Gaskill
THE WORLD TRADE ORGANIZATION Amrita Narlikar
WORLD WAR II Gerhard L. Weinberg

Kenneth Falconer

FRACTALS

A Very Short Introduction

OXFORD
UNIVERSITY PRESS

Great Clarendon Street, Oxford, OX2 6DP,
United Kingdom

Oxford University Press is a department of the University of Oxford.
It furthers the University's objective of excellence in research, scholarship,
and education by publishing worldwide. Oxford is a registered trade mark of
Oxford University Press in the UK and in certain other countries

First Edition published in 2013

Published in the United States of America by Oxford University Press
198 Madison Avenue, New York, NY 10016, United States of America

British Library Cataloguing in Publication Data
Data available

ISBN 978-0-19-967598-2

Printed and bound by
CPI Group (UK) Ltd, Croydon, CR0 4YY

Contents

Preface

To many people, the word 'geometry' conjures up circles, cubes, cylinders, and other regular or smooth objects. Familiar artefacts, such as buildings, furniture, or cars, make wide use of such shapes. However, many phenomena in nature and science are anything but regular or smooth. For example, a natural landscape may include bushes, trees, rugged mountains, and clouds, which are far too intricate to be represented by classical geometric shapes.

Surprisingly, apparently complex and irregular objects can often be described in remarkably simple terms. Fractal geometry provides a framework in which a simple process, involving a basic operation repeated many times, can give rise to a highly irregular result. Fractal constructions can represent natural objects but also give rise to a vast array of other shapes, which may be of extraordinary complexity. The phrase 'the beauty of fractals' is often heard, a phrase which reflects the unending intricacy of fractal designs alongside the simplicity which underlies their ever-repeating form. Indeed, complex but attractive fractal pictures have become an art form in their own right, with exhibitions, competitions, and their use on designer clothing.

The aim of this book is to show, in basic terms, how fractals may be constructed, described, and analysed as geometrical objects and how these 'mathematical' fractals relate to the 'real' fractals of nature or science. Geometry, particularly fractal geometry, is very much a visual subject, and diagrams and a visual intuition are key to its appreciation. Inevitably some mathematics is required, but this is presented alongside a visual and intuitive interpretation and hopefully will not present too great an obstacle to the dedicated reader prepared to think through the concepts involved. In a couple of places, to keep the mathematics in the main part of the chapters to the minimum, further details are deferred to an Appendix at the end of the book.

Since fractals became popular in the 1980s, their theory and applications have developed beyond recognition, with fractals pervading many areas of mathematics and science, as well as economics and social science. Fractals are central to a great deal of sophisticated research, which has at its heart the simple ideas that are sketched in this Very Short Introduction. If the reader is able to appreciate some of these ideas then the book will have succeeded in its aim.

I am most grateful to Isobel Falconer, Timothy Gowers, and Emma Ma for reading earlier drafts of this book, to Ben Falconer and Jonathan Fraser for assistance with producing some of the figures, and to Carol Carnegie, Prabhavathy Parthiban, Joy Mellor, and Latha Menon of Oxford University Press for their work in seeing this Very Short Introduction through to publication.

Kenneth Falconer

St Andrews, Scotland, 2013

List of illustrations

Chapter 1
The fractal concept

The rise of fractals

Since ancient times, mathematics and science have developed alongside each other, with mathematics used to describe, and often explain, observed natural and physical phenomena. In many areas this marriage has been highly successful, indeed much of what we enjoy in modern life is a consequence of its success. For example, the mathematical methods and laws introduced by Isaac Newton underlie the operation of almost everything mechanical, from riding a bicycle to the orbit of a spacecraft. James Clark Maxwell's equations of electromagnetism hastened the understanding and development of radio communication. A picture on a computer screen can only be created or moved around using a mouse because a great deal of geometrical calculation has gone into designing the software.

Nevertheless, there are many phenomena which, although governed by the basic laws of science, were historically regarded as too irregular or complex to be described or analysed using traditional mathematics. Classical geometry concentrated on smooth or regular objects such as circles, ellipses, cubes, or cones. The calculus, introduced by Newton and Leibniz in the second half of the 17th century, was an ideal tool for analysing smooth objects and rapidly became so central both in mathematics and science that any attempt to consider irregular objects was sidelined. Indeed, many natural phenomena were overlooked, perhaps deliberately, because their irregularity and complexity made them difficult to describe in a form that was mathematically manageable.

It was not until the late 1960s that the study of irregular figures started to develop in a systematic way, largely as a result of efforts by the French-American polymath Benoit Mandelbrot (1924–2010), often referred to as 'the Father of Fractals'. In his 1982 book *The Fractal Geometry of Nature* he wrote: 'Clouds are not spheres, mountains are not cones, coastlines are not circles, and bark is not smooth, nor does lightning travel in a straight line.' He argued that highly irregular objects should be regarded as commonplace, rather than as exceptional, and moreover that many phenomena from across physics, biology, finance, and mathematics have irregularities of a form that are akin to each other. Mandelbrot introduced the word *fractal* as a general description for a large class of irregular objects, and highlighted the need for a fractal mathematics to be developed, or in some cases retrieved from isolated forgotten papers.

Since the 1980s fractals have attracted widespread interest. Virtually every area of science has been examined from a fractal viewpoint, with 'fractal geometry' becoming a major area of mathematics, both as a subject of interest in its own right and as a tool for a wide range of applications. Fractals have also achieved a popular vogue, with attractive, highly coloured, fractal pictures appearing in magazines, books, and art exhibitions, and even used for scenery in science fiction films. Further public interest has been generated with the widespread use of computers at home and at school, enabling anyone with a basic knowledge of programming to generate intricate fractal pictures by repeatedly applying a simple operation.

Of course, there is always a difference between idealized mathematical objects and the real phenomena that they may represent. A circle has a precise definition as those points on a piece of paper or other flat surface whose distance from a centre exactly equals a given radius, and a sphere consists of those points in space with the same property. We may refer to a coin or wheel as circular, and an orange or the earth as spherical, but these are only approximate descriptions. On close examination, the surface of an orange is dimpled and may be slightly flattened at the top and bottom so it is not quite a sphere, and the earth's surface is covered with hills and valleys. Nevertheless, in the right context it is extremely useful to regard these objects as circles or spheres. If you want to calculate the number of oranges that can be packed into a box it is good enough to assume that they are spherical, and when computing orbits of the earth around the sun or the moon around the earth little is lost by assuming that these bodies are indeed spheres.

In the same way, we will define fractals in a mathematically exact manner. But we will also encounter natural, physical, and economic phenomena that can usefully be regarded as fractals when viewed over an appropriate range of scales. These will be thought of as 'real', as opposed to 'mathematical', fractals, with the fractal description inevitably ceasing to be valid if they are examined too closely. This book contains pictures of various fractals, but they are no more than approximations to the exact mathematical objects which possess detail at a far finer scale than anything that can be printed on a page.

A first fractal construction—the von Koch curve

Let's start with a shape that can be drawn, at least very roughly, using just a pencil and eraser.

Take a straight line and divide it into three equal pieces. Erase the middle piece, and replace this by the other two sides of an equilateral (i.e. equal-sided) triangle on the same base. This gives a chain of four shorter, joined-up, straight-line segments: see

Figure 1. Now do exactly the same thing with each of these four pieces: remove the middle thirds and replace by the other two sides of the equilateral triangles on the same base, to get a chain of 16 straight pieces. And now repeat this process again and again. The stages of this construction (which of course have been drawn on a computer rather than by hand) are shown in Figure 2(a). In principle we carry on forever with this procedure, but before long the figure becomes indistinguishable to the eye from the 'curve' labelled F. This is known as the *von Koch curve* introduced and studied by the Swedish mathematician Helge von Koch (1870–1924) in 1906. (Note that the word *curve* is used here simply to mean a path that can be traced from one end to the other, without any implication of smoothness.)

Look more closely at the von Koch curve. On zooming in on the curve, however much we magnify it, irregularities in its shape will always be apparent—indeed the curve contains tiny von Koch curves just as wiggly as the original, see Figure 3. This is a direct consequence of the construction wherein the very small line segments were treated in just the same way as the original one but at a smaller size. An object which has such irregularity at arbitrarily small scales is said to have a *fine structure*. This is very different from, say, a circle where a small portion of the perimeter sufficiently magnified will appear almost indistinguishable from a straight line.

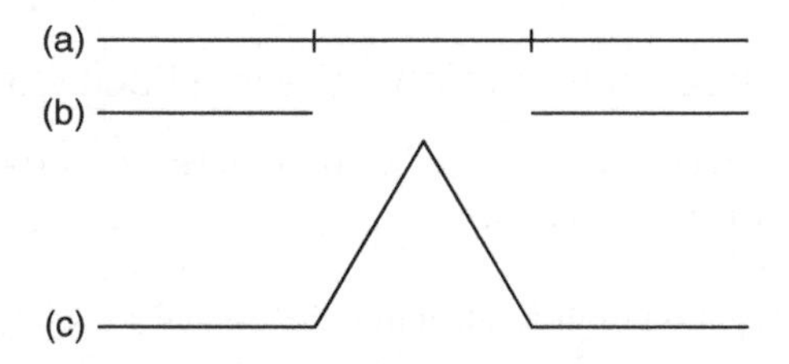

1. The basic step in constructing the von Koch curve: (a) divide a line into three equal parts, (b) remove the middle part, (c) replace the missing part by the other two sides of an equilateral triangle

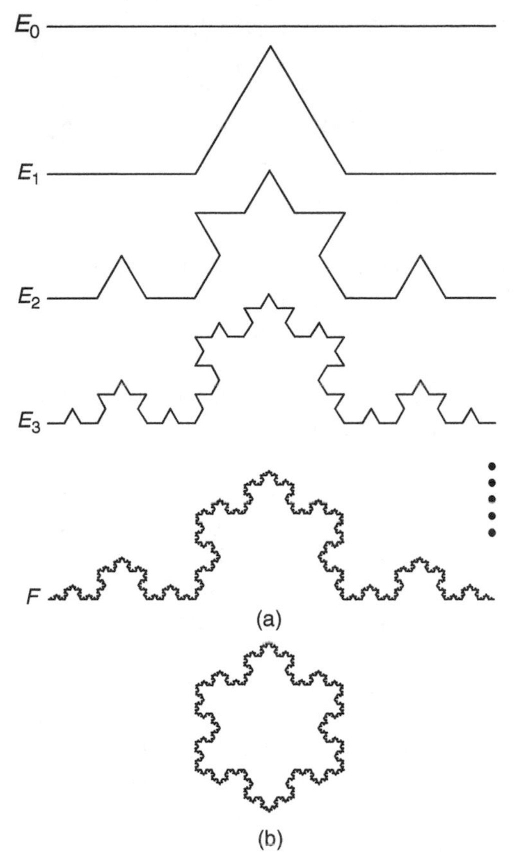

2. (a) The first few stages $E_0, E_1, E_2, \ldots$ of construction of the von Koch curve F, (b) three such curves joined together to form the von Koch snowflake

Looking from another view point, the von Koch curve is *self-similar*, that is it is made up of smaller scale copies of itself. In particular we could cut it into 4 parts each a ⅓ scale copy of the entire curve. Thinking of this another way, 4 photocopies of the whole von Koch curve at 33⅓ per cent

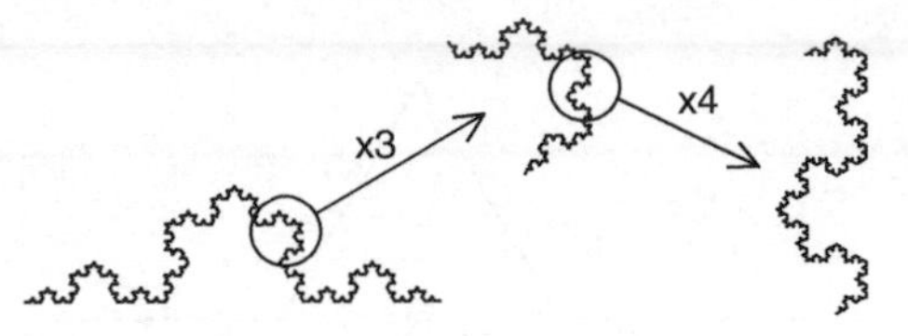

3. Repeated magnification reveals more irregularity

reduction can be positioned and joined together to form a curve identical to the original. But there are also self-similarities at smaller scales: the curve is made up of 16 copies of itself at $\frac{1}{9}$ scale, 64 copies at $\frac{1}{27}$ scale and so on—everywhere on the curve and at whatever enlargement there are many tiny copies of the whole curve.

Smooth curves—circles, ellipses, etc.—have a tangent at each point, that is a straight line that neatly touches the curve. This can be thought of as the instantaneous direction of travel of a point traversing the curve. This notion of a tangent or direction at each point is central to way that mathematicians have for centuries studied curves (it is the essence of the 'calculus'). However, the von Koch curve does not have a well-defined direction or slope at any point and one cannot draw tangents to the curve. It is far too irregular to be described in traditional geometrical language and, unlike classical shapes, cannot be expressed by a 'simple' formula—*the methods of classical mathematics are not applicable* to the von Koch curve.

To find the perimeter length of a circle one might 'walk' around the circle taking very small steps, and multiply the step length by the number of steps taken. If the circle is of radius 1 (kilometres, metres, yards—the units don't matter so long as we are consistent), then if the steps are small the answer will be very close to 6.283 (that is 2π) which is the perimeter length. We might try the same approach to measure the length of the von Koch curve. Assuming that the initial line segment in the construction has length 1 then traversing the curve with a step of length $\frac{1}{10}$ will require about 19 steps, corresponding to a length walked of $19 \times \frac{1}{10} = 1.9$. If we reduce our step lengths to $\frac{1}{100}$, then our walk around the curve will visit many more of the small 'corners' of the curve and will take about 334 steps, a length $334 \times \frac{1}{100} = 3.34$. And if we take tiny steps of, say, length $\frac{1}{1,000,000}$ (one-millionth), following around the irregularities of the curve will necessitate about 37.25 million steps, giving a length of 37.25. Unlike with the circle, trying to measure the von Koch curve by dividing it into shorter and shorter steps just gives ever larger estimates for its length. *The size depends on the scale at which the length is measured*—another property that distinguishes the von Koch curve from the circle or other classical geometrical figures.

In all these ways the von Koch curve is a very intricate and complicated object. Yet, in other ways it is very simple. Its construction can be described in a short sentence: 'Repeatedly replace the middle third of each line segment with the other two sides of an equilateral triangle.' It is a *recursive construction*—that is it involves performing a simple step over and over again.

If we take three copies of the von Koch curve and join them corner to corner in a 'triangle', we get a rather idealized 'von Koch snowflake', shown in Figure 2(b)—with a bit of imagination the curve has a *natural appearance.*

To summarize, the von Koch curve possesses the following properties:

- *Fine structure*—detail at all scales, however small
- *Self-similarity*—made up of small scale copies of itself in some way
- *Classical methods of geometry and mathematics are not applicable*
- *'Size' depends on the scale at which it is measured*
- *A simple recursive construction*
- *A natural appearance*

A curve or other object with such properties is called a *fractal*, a word coined by Benoit Mandelbrot in 1975 from the Latin *fractus*, meaning 'broken'. These features will be prominent in the wide range of fractals that will be encountered in the following pages.

A word of caution about this 'definition' of a fractal. We have termed an object a fractal if it satisfies several conditions, some of which are a little vaguely framed. Is there a more precise definition of a fractal? There has been considerable debate about this ever since the term was introduced. Mandelbrot originally proposed a technical definition in terms of dimensions, but this was abandoned because there were plenty of objects that did not fit in with the definition but which clearly ought to be considered fractals. The current consensus is to regard something as a fractal if all or most of the properties listed above (along with one or two other more technical ones) hold in some form. This is somewhat analogous to the way biologists define 'life'. Something is held to be alive if it has all or most of the characteristics on a list: ability to grow, ability to reproduce, ability to respond to stimuli in some way, etc. Nevertheless, there are things which are obviously 'alive' but which don't have quite all the properties on the list.

Some more examples

Fractals with a completely different appearance may be constructed by a very similar recursive procedure. We obtained the von Koch curve by repeatedly replacing each straight line segment by a simple figure _/_ sometimes called a *generator* or *motif* of the curve. Figure 4 shows some fractals constructed using other generators, again by replacing line segments by scaled down copies of the generator. The first example is a 'squig' curve generated by a 'square wave'. The second is a fractal 'grass' with its natural appearance a consequence of the 'twigs' of the generator being at a slight angle to the main 'stem'. Note how the list of fractal properties applies in each case. There are endless

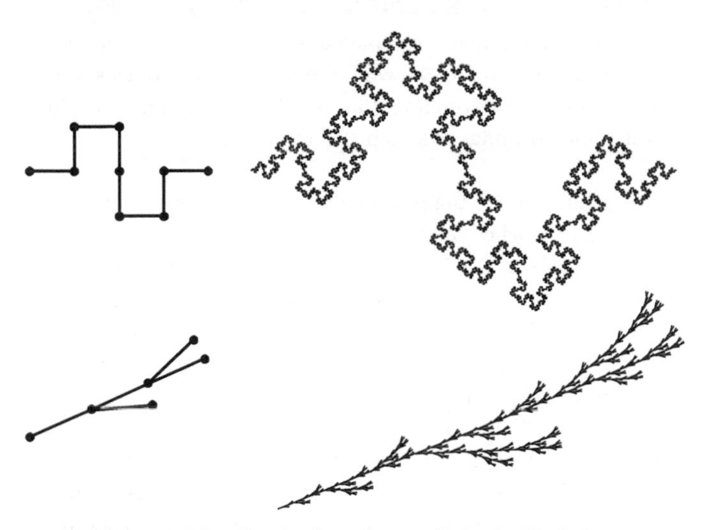

4. A squig curve (above) and a fractal grass (below) with their generators: the line segments in the generators are repeatedly replaced by scaled down copies to form the fractals

possibilities for producing fractals in this way from different generators.

Coordinates, functions, and itineraries

The fractals we have seen are drawn on a piece of paper and could equally well be displayed on a computer screen. Such a flat surface is referred to as a *plane*. Of central interest to us are black and white 'pictures' drawn on a plane, that is any collection of points on the plane regarded as an entity. We will refer to such a collection of points as an *object*, a *shape*, a *figure*, or, to use the mathematician's term, a *set*. Thus, in general, a 'set' might refer to a circle, a von Koch curve, a collection of paint splashes on a page, or the silhouette of a person.

To describe fractals, and indeed any set, we need a means of specifying positions on a plane. The standard way of doing this uses coordinates, sometimes called Cartesian coordinates after René Descartes (1596–1650) who introduced them, which specify points in a similar way to longitude and latitude on a map, or the eastings and northings of a map grid reference.

Choose some point of the plane, called the *origin*. Any point of the plane may be reached from the origin by travelling a certain distance horizontally and then turning through a right angle and travelling another distance vertically. The location of the point is given by a pair of numbers, the first being the horizontal distance travelled and the second the vertical distance. These numbers are the *coordinates* of the point. Thus the pair of coordinates (3, 2) gives the position of the point 'distance 3 along and 2 up', see Figure 5. It is usual to draw horizontal and vertical lines through the origin, called the x-*axis* and y-*axis*, respectively, which may be marked with scales that indicate the distances in each of the perpendicular directions. By convention, positive numbers represent distances to the right or upwards from the origin, and negative numbers denote distances to the left or downwards. We usually identify the coordinate pair with the point itself, so we refer to 'the point (3, 2) in the plane'.

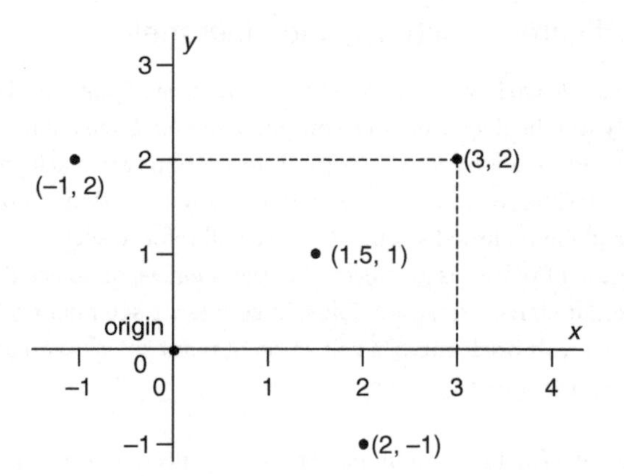

5. Points in the plane with their coordinates

We also need to relate points in the plane, and one very useful way of doing this is using 'functions'. For our purposes, a *function* is simply a rule or formula which, for each point in the plane, specifies some other point. We will frequently think of a function as an instruction that tells you how to move around the plane: if you are at some point, the function tells you another location to which to move. (Technically, such a function should be referred to as a 'function from the plane to itself', but all functions referred to in this book will be of this form.) We will use arrows to denote functions, for example, the function

$$(x,y) \to (x+1,y) \qquad (1)$$

says that if you are at the point with coodinates (x, y) you move to the point with coordinates $(x + 1, y)$. The effect of this function on a specific point, (2, 5), say, is found by substituting the pair (2, 5) into the formula (1) replacing x by 2 and y by 5, so that $(2,5) \to (2 + 1, 5) = (3, 5)$; thus the function moves the point (2,5) to (3,5). Similarly, it moves the point (3, 7) to (4, 7) and (−3, −7) to (−2, −7). This function has the effect of adding 1 to the first coordinate of a point and leaving the second coordinate unchanged, so geometrically the function moves every point a distance 1 to the right.

For another example, the function given by

$$(x,y) \to (\tfrac{1}{2}x, \tfrac{1}{2}y) \qquad (2)$$

halves both coordinates, so takes each point to the point midway between it and the origin, so, for example, $(6, 2) \to (3, 1)$ and $(3, 2) \to (1.5, 1)$, see Figure 5.

We saw with the von Koch curve that repeating a simple operation over and over again produced an intricate fractal. Similarly, applying a function repeatedly can also lead to highly complex objects. Given a function and some initial point in the plane the function tells us to move to some new point. Applying the function to this second point takes us to a third point, applying the function to this third point takes us to a fourth point, and so on. Repeatedly applying the function takes us on a tour or *itinerary* around the plane, visiting a sequence of points in turn. This is rather like a treasure hunt of the type popular at children's parties. An initial clue tells a child to go to some location where another clue may be found. This directs them to a further clue, and so on, so that they visit an itinerary of locations around a garden or park. A function is just a concise way of expressing such 'clues'—applying the function to each point reached gives the next point to be visited. The process of repeatedly applying the function is known as *iteration* or *iterating the function* and the points visited are called the *iterates* of the initial point.

We can work out itineraries for the function (1) above by repeatedly adding 1 to the first coordinate. The itinerary starting at the point (2, 5) is

$$(2,5) \to (3,5) \to (4,5) \to (5,5) \to (6,5) \to (7,5) \to (8,5) \to \ldots$$

where the arrows indicate the movement of the iterates across the plane and the dots indicate that we continue in the same way indefinitely (see Figure 6). Of course the itinerary depends on the starting point: if we started at (–2,–1) the itinerary under (1) would be

$$(-2,-1) \to (-1,-1) \to (0,-1) \to (1,-1) \to (2,-1) \to (3,-1) \to \ldots$$

Often of particular interest is what happens to the itinerary in the long term. For the function (1), for either of these initial points, and indeed for all starting points, the itineraries go further and further away to the right, never to return.

If we iterate the function (2) the behaviour is quite different. Each application of the function halves both coordinates, so starting at (8, 4) gives the sequence

$$(8,4) \to (4,2) \to (2,1) \to (1,0.5) \to (0.5,0.25) \to (0.25,0.125) \to (0.125,0.0625) \to \ldots$$

an itinerary which rapidly approaches the origin (0, 0), see Figure 6. Indeed iteration under (2) gives an itinerary that approaches (0, 0) whatever the initial point.

Fractals by iteration

For certain choices of function, the itineraries can behave in a much more complicated manner. The *Hénon function*, introduced by Michael Hénon in 1976, is given by

$$(x, y) \to (y + 1 - 1.4x^2,\ 0.3x)$$

(the precise form of this formula is irrelevant here, but it is the introduction of the term involving x^2, that is x-squared or $x \times x$,

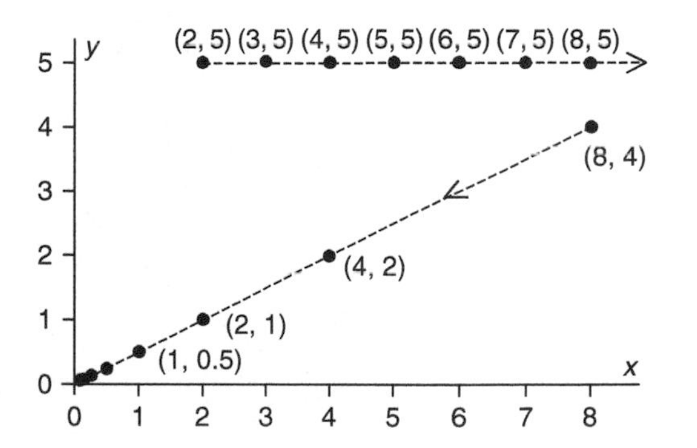

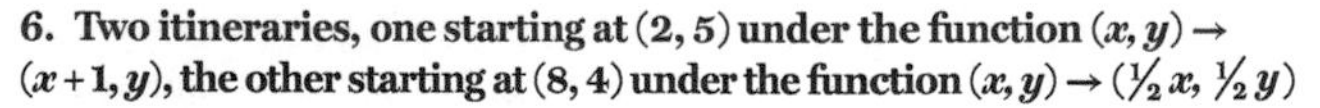
6. Two itineraries, one starting at (2, 5) under the function $(x, y) \to (x + 1, y)$, the other starting at (8, 4) under the function $(x, y) \to (\frac{1}{2}x, \frac{1}{2}y)$

that complicates things enormously). Computing a few iterates, starting at (1,1) say, gives

$$(1,1) \to (0.6, 0.3) \to (0.796, 0.18) \to (0.293, 0.239)$$
$$\to (1.119, 0.088) \to (-0.664, 0.336) \to \ldots$$

with the itinerary showing very little overall pattern, so we need to look at many more iterates to get a feel for what is happening. The first 20,000 or so points of an itinerary obtained by iterating the Hénon function are plotted in Figure 7. Ignoring the first few points (a 'settling down' period), the sequence of iterates jumps about in a seemingly random manner but nevertheless builds up a curved, stratified figure called the *attractor* of the function—the points of the itinerary are 'attracted' to it. Moreover, the attractor obtained does not depend on the starting point of the iteration (provided it is not too far away from the origin).

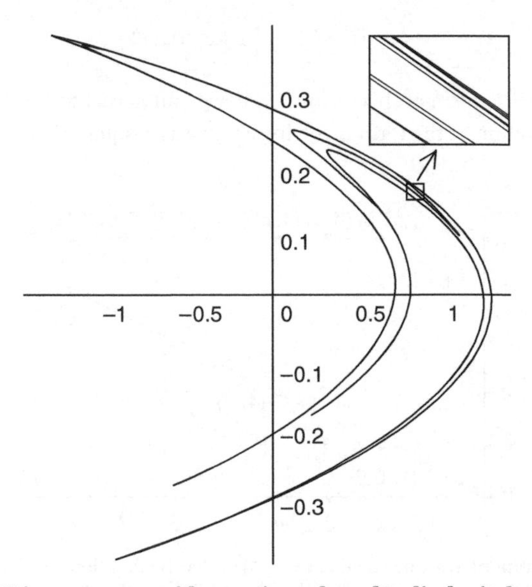

7. The Hénon attractor with a portion enlarged to display its banded structure

The structure of this attractor merits closer examination. On magnifying a small portion, the curves apparently forming the attractor are seen to comprise many closely spaced (almost) parallel lines. Further magnification shows that these lines themselves are made up of many more closely spaced lines, and so on—the attractor has a fine structure formed by nearly parallel lines. Thus iteration of the Hénon function yields a fractal attractor, sometimes called a *strange* attractor. Although the attractor is highly intricate, it is completely defined by the Hénon function given by a formula of just a few symbols. Once again, repeated application of a simple operation gives rise to a complex fractal form.

A nice feature of iteration is that it is very easy to realize on a computer. Computers are very good at performing the same operation over and over again, and a few lines of code are enough to calculate the itinerary of a function from a given starting point, with each point obtained by applying the function to its predecessor, and this may be plotted on a screen to give a picture of the attractor. It is easy to enlarge parts of the attractor so that any fractal structure can be investigated visually. This contrasts markedly with mathematical analysis of many attractors. Even for simple functions, explanations of why an attractor has a particular fractal structure can be beyond the reach of current mathematics.

What can be done with fractals?

It is natural to query the role of highly irregular fractal objects within mathematics and science. The clue to this comes from classical geometry—the questions that have been addressed for centuries concerning traditional geometrical shapes are precisely those that more recently have come to be asked about fractals.

- *Description* A circle consists of those points that lie at some constant distance from a given centre, and an ellipse comprises the points such that the sum of their distances from two fixed points (the 'focii') is constant. Are there such concise descriptions of fractals? We will see how simple 'templates' can encode intricate fractal forms.

- *Measurement* Almost the first thing one asks about any mathematical object is 'how big is it?'. For a circle or rectangle we can measure the area or the perimeter length. When considering the size of a fractal the notion of 'dimension' turns out to be central.
- *Geometrical properties* If a circular ring is held up under a light or in the sun, the shadow will be an ellipse, which may be highly elongated or almost circular, depending on its angle to the light. Do fractals have any such geometrical properties?
- *Occurrence in other areas of mathematics* Geometrical shapes provide solutions to many mathematical problems: the circle is the curve of given length that encloses greatest area; the parabola is the figure such that all parallel rays are reflected to pass through a single point—the focus. Are there mathematical questions to which fractals provide the answers?
- *Applications to science and social science* Classical geometry answers many important questions in science and other areas: planets follow (more or less) elliptical orbits; a thrown stone traces a parabolic path, atoms in crystals form regular lattice patterns, DNA molecules form a double helix, etc. When can fractals provide the solutions to practical problems?

Such natural questions will be discussed in subsequent chapters.

Chapter 2
Self-similarity

Self-similar fractals and their templates

The concept of similarity features prominently in classical geometry, notably in the writings of the Greek mathematician Euclid (*c.*300 BC), the 'Father of Geometry', who laid the foundations of rigorous geometrical argument. Two figures in the plane are *similar* if they have the same shape, but not necessarily the same size, so that one may be obtained from the other first by scaling and then sliding it around and rotating it, perhaps flipping it over. In modern parlance, two objects in the plane are similar if it is possible to make a reduced or enlarged photocopy of one and position it to coincide exactly with the other, perhaps turning it over first. If this can be done without scaling, in other words, with a same size photocopy, the objects are called *congruent*. Otherwise, the reduction or enlargement factor required to produce the similar copy is called the *scale* or *scaling ratio* of the copy, which may be expressed as either a fraction or percentage. Thus if a figure is a ¼ scale (or 25 per cent scale) copy of another then all the lines in the copy are ¼ of the length of the corresponding lines in the original.

All circles are similar to each other, as are all squares. However, two triangles are similar precisely when the three angles of one triangle are the same as the three angles of the other. Two

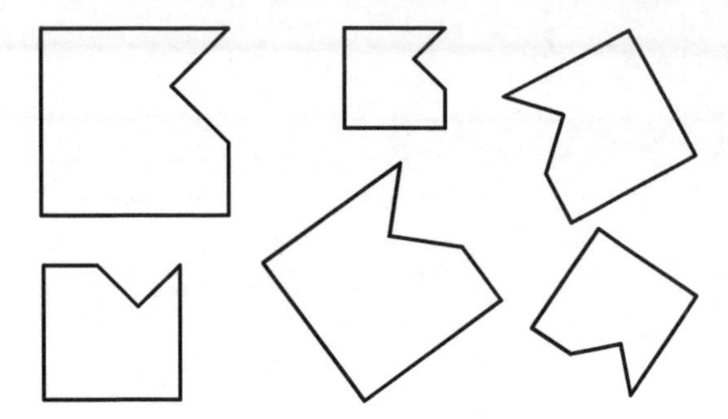

8. The figures shown are all similar to each other

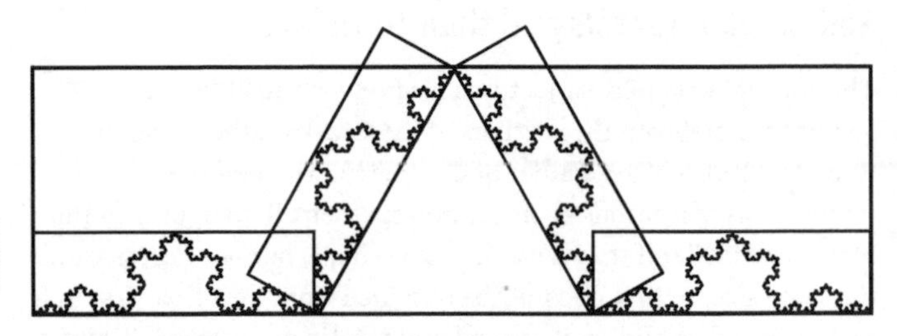

9. The von Koch curve together with its defining template

rectangles are similar just when the length of the longer side divided by that of the shorter side is the same for both rectangles. Figure 8 shows several figures that are all similar to each other.

A *self-similar* set is one that is made up of several smaller similar copies of itself. We've seen that the von Koch curve is self-similar since it comprises 4 suitably placed scale ⅓ copies of itself. This may be represented diagrammatically by a *template* consisting of one large rectangle and 4 smaller ones, each a ⅓ scale copy of the large one, see Figure 9. The part of the von Koch curve framed by each of the smaller rectangles is a ⅓ scale copy of the whole curve framed by the large rectangle. The size and position of the rectangles specify the scaling and positioning needed to fit together the four smaller copies to make up the whole von Koch curve.

But much more than this is true. The template is a simple diagram made up of classical geometrical shapes, namely five rectangles—there is nothing fractal about it. Nevertheless, the template completely defines the fractal: the von Koch curve is (essentially) the *only* object that is made up of smaller scaled copies of itself positioned as indicated by the template. It is easy to reconstruct the von Koch curve from the template by repeated substitution of a scaled template into the smaller rectangles. The first couple of stages are indicated in Figure 10 from which it

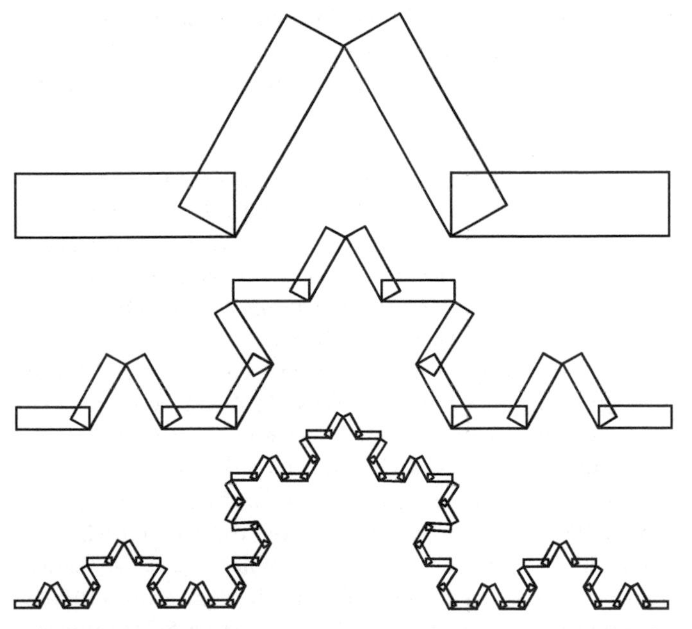

10. Reconstruction of the von Koch curve by repeated substitution of the template in itself

should be clear that the small rectangles form patterns that get progressively closer to the von Koch curve itself.

This is a particular case of a very general and powerful method of describing and constructing fractals. A *template* consists of some simple shape (square, rectangle, triangle, etc.) and a number of smaller similar copies of the shape positioned somehow in the plane. Each of the smaller shapes represents a *similarity transformation* or *scaling transformation* that takes the larger shape and scales and repositions it to coincide with the smaller copy, scaling and positioning any figure drawn inside correspondingly. A fundamental property is that templates define fractals. That is, given a template, there is essentially just one figure, usually a fractal, made up of smaller copies of itself scaled and positioned according to the shapes in the template. (This statement is not quite accurate, but we will be more precise soon.) This figure is sometimes called the *attractor* of the template.

For a template made up of similar shapes, the attractor will be *self-similar*. Figures 11 and 12 show several self-similar fractals with their templates. In each case note that what is seen in each of the smaller regions of the template is a scaled copy of the whole. Figure 11 is known as the *(right-angled) Sierpiński triangle* made up of 3 copies of itself at scale ½. The *snowflake* of Figure 12(a) is made up of 4 smaller copies of itself; 3 are at scale ¼ whilst the larger central copy is at scale ¾ and has been rotated by 180° (or alternatively reflected in a horizontal line). This example illustrates two variants: the scalings of the similarity transformations do not all have to be the same, and some of the copies may be reflected or rotated. The *spiral* in Figure 12(b) is remarkable in that it is defined by just 2 similarity transformations. The first scales down only slightly with a 19/20 ratio but with a 45° rotation, and the second transforms the whole picture to the spiral at the end of one of the arms of the main spiral scaling at 1/5.

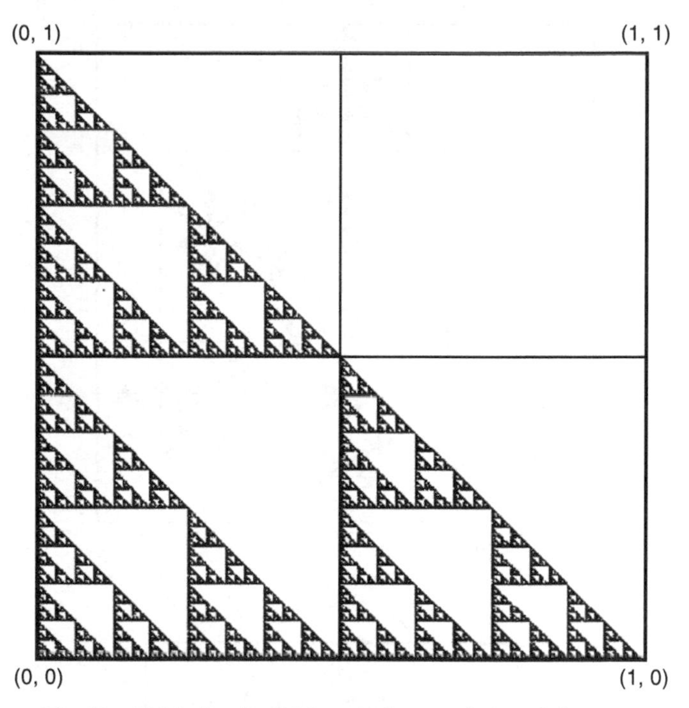

11. The Sierpiński triangle with its template consisting of a large square of side 1 and three squares of side ½

Orientation

In saying that a template completely defines a fractal we have glossed over the possibility that if the shape in the template has some symmetry then there will be more than one similarity that transforms the shape to a smaller copy. For a simple instance, there are 8 different similarities that transform a square to a smaller square. The scaling can be direct or there can be a rotation through 90°, 180°, or 270°, or a 'mirror' reflection in a vertical, horizontal, or one of the two diagonal lines, see Figure 13; we refer to these possibilities as the *orientations* of the scalings. Thus for the template used for the Sierpiński triangle in Figure 11 we have

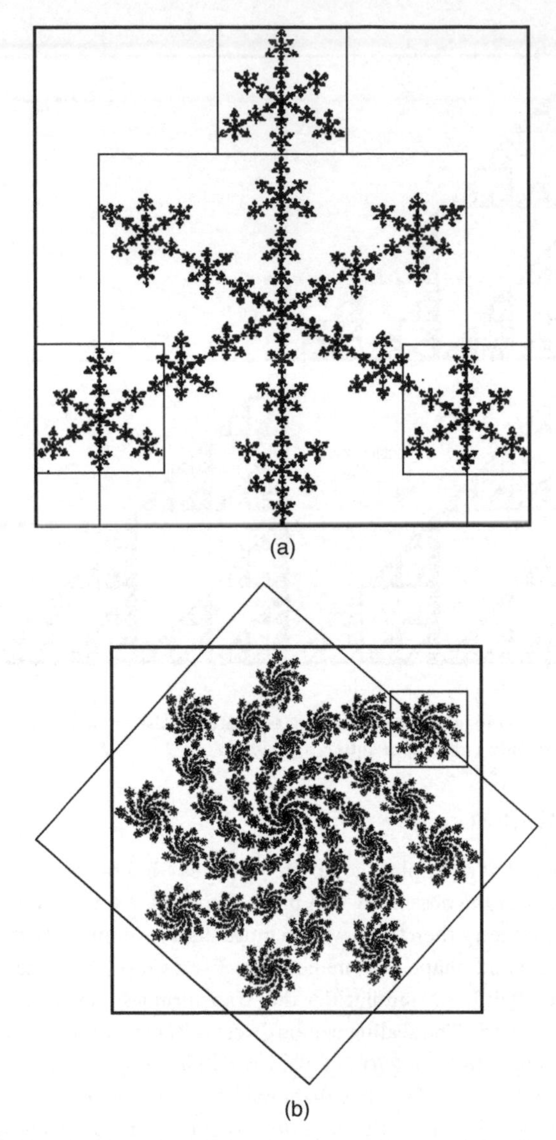

12. Self-similar fractals with their templates: (a) a snowflake, (b) a spiral

13. The 8 symmetries of the square indicated by the positions of the face

a choice of 8 similarity transformations that move the large square onto each of the 3 smaller squares, so there are $8 \times 8 \times 8 = 512$ different ways of choosing the three similarity transformations. Figure 14 shows two of the possible attractors resulting from the same template but with different choices of orientation for the similar copies. Note that the attractor of Figure 14(a) may be obtained from 8 different combinations of orientation of the similarities: the portion of the attractor in each of the three small squares may be obtained from the whole attractor either with a rotation or by a reflection in a diagonal. On the other hand, there is only one choice of orientation for each of the three similarities that gives the attractor of Figure 14(b).

It turns out that there are 456 different attractors defined by this 'three squares' template. Of these, 8 have mirror symmetry about the rising diagonal, and all of these result from 8 different combinations of orientations for the three small images. The other 448 result from a unique combination of orientations.

Thus if the shapes in a template shape have some symmetry, the template alone need not define a unique attractor; however, once the orientations of the similarity transformations are specified,

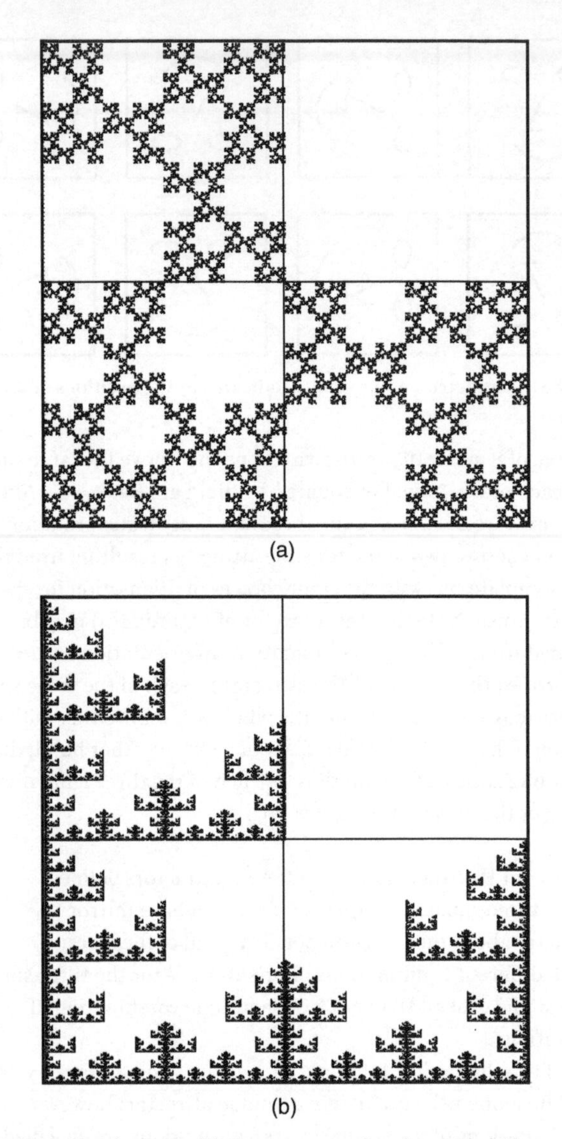

14. Self-similar fractals with the same template but with different orientations of the similar copies

there ceases to be any ambiguity. This important fact may be stated as follows:

> Given a template, there exists a unique set (called the attractor and usually a fractal) that is made up of similar copies of itself which are scaled and positioned according to the shapes in the template, provided that the orientation (i.e. reflections or rotations) of each copy is specified.

(The perceptive reader may realize that this is still not strictly true—there are 'pathological' sets which also fit the templates: for example the 'empty set', which has a blank picture, fits any template. But for here this statement is good enough.)

Templates and functions

Functions were introduced as a rule or formula that takes each point of the plane and 'moves' it to a new position. There is another way of thinking of functions which emphasizes their geometrical effect. A (black and white) picture in the plane is made up of a large collection of points (these might be thought of as the dots or pixels on a computer screen that make up the picture). A function moves each individual point to a new point. Thus the *set* of points that make up the picture are moved by the function to a *set* of new positions, which taken together form a new picture. The function *transforms* the original set or picture into a new one, Figure 15. The word *transformation* is used synonymously with 'function', particularly when thinking about the geometrical effect on objects. (The words *map* or *mapping* are also used—in the familiar sense of the word, a map is really a function that associates each point on the ground with a point on the page of an atlas, at the same time transforming a coastline, say, to a corresponding wiggly curve on the page.)

We have encountered the function $(x, y) \rightarrow (\frac{1}{2}x, \frac{1}{2}y)$ which moves each point in the plane to the point midway between it and

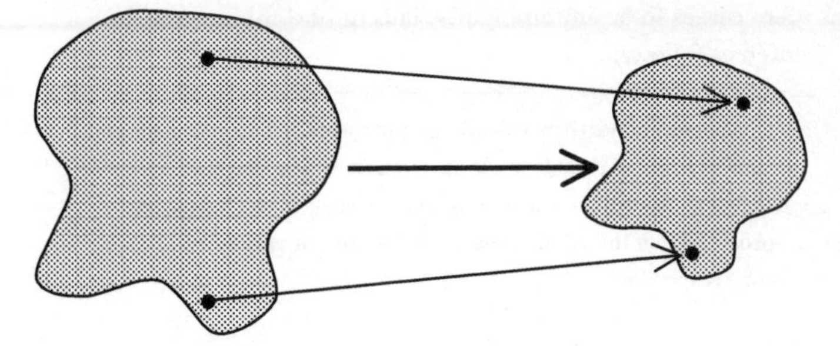

15. A function moves points to points and so transforms sets of points to sets of points

the origin. Its geometrical effect is to shrink everything down towards the origin by a factor ½. This function will transform any picture to a ½ size copy—thus it represents a similarity transformation of scale ½.

Templates are simply a diagrammatic way of expressing a family of transformations or functions. Each of the smaller shapes codes the similarity transformation which takes the large shape and scales it down to the size and position of the smaller one. Each point inside the large shape is moved to the corresponding position within the smaller one, so any picture inside the large shape is transformed to a similar picture inside the smaller shape.

The Sierpiński triangle of Figure 11 has a template comprising the 'unit square', that is the square of side length 1 with corner with coordinates (0, 0), (1, 0), (0, 1), (1, 1), and three smaller squares of side-lengths ½. The three functions giving the three similarity transformations that transform the unit square to the bottom left, bottom right, and top left squares respectively, have the formulae:

$$(x, y) \to (\tfrac{1}{2}x, \tfrac{1}{2}y); \qquad (x, y) \to (\tfrac{1}{2}x + \tfrac{1}{2}, \tfrac{1}{2}y); \qquad (x, y) \to (\tfrac{1}{2}x, \tfrac{1}{2}y + \tfrac{1}{2}). \tag{1}$$

The first function just shrinks everything down towards the origin by a scale factor $\frac{1}{2}$. The second function does the same thing and then, because of the '+ $\frac{1}{2}$' term, shifts everything a distance $\frac{1}{2}$ to the right. The third function scales by a factor $\frac{1}{2}$ and then shifts a distance $\frac{1}{2}$ up. As well as transforming the square, the functions transform the fractal itself. Thus $(x, y) \to (\frac{1}{2}x, \frac{1}{2}y)$ shrinks the entire Sierpiński triangle onto the scale $\frac{1}{2}$ copy inside the bottom left square. Similarly, the other two functions in (A) transform the entire Sierpiński triangle onto the similar copies in the bottom right and top left small squares. Together, these three scale $\frac{1}{2}$ copies make up the entire Sierpiński triangle. The three functions (1) describe the basic self-similarities of the Sierpiński triangle—they are a mathematical way of describing the template. The Sierpiński triangle is completely defined by the template and orientations, or equivalently by the three functions. However, the functions have the advantage that they also encode the orientations of the transformations—these do not have to be noted separately. For example, the three functions that transform the fractal in Figure 14(b) to the similar copies in the three smaller squares are

$$(x,y) \to (\tfrac{1}{2}x, \tfrac{1}{2}y); \qquad (x,y) \to (-\tfrac{1}{2}x+1, \tfrac{1}{2}y);$$
$$(x,y) \to (\tfrac{1}{2}x, \tfrac{1}{2}y+\tfrac{1}{2}),$$

with the second function differing from that in (1).

We saw that a template (with given orientations) defines a unique fractal. This fact can be expressed in the language of transformations or functions, namely that a family of transformations defines a unique object.

> Given a family of contracting transformations, there exists a unique set (called the *attractor* and usually a fractal) such that, taken together, the transformed copies of the set make up the entire set.

This 'theorem' or 'mathematical fact' can be stated and proved in a formal mathematical way. Note that the theorem states two things: that such a set exists, and that it is unique—there is only one figure which satisfies the stated property. This is typical of many major theorems in mathematics which guarantee the existence and uniqueness of entities with certain properties. Notice also that this statement does not require the transformations to be similarities—it holds for any *contracting* transformations, that is transformations that shrink sets in some way, and similarities of scale less than 1 are a special case of this. A family of contracting transformations is called an *iterated function system*, and the theorem is sometimes called the 'Fundamental Theorem of Iterated Function Systems'. In these terms the theorem simply asserts that an iterated function system has a unique attractor.

Drawing fractals

Given a template or a family of transformations, how can we obtain good pictures, usually on a computer, of the unique fractal so defined? One widely-applicable method was indicated above for the von Koch curve in Figure 10 where we repeatedly substituted scaled down copies of the template into itself to get increasingly good approximations to the fractal.

An alternative method that is easy and efficient to program, called the *Chaos Game*, was introduced by Michael Barnsley. This produces a sequence of points in the plane (i.e. on a computer screen) that give a very good approximation to the fractal. This is rather like drawing fractals by iterating a function as we did for the Hénon attractor, the difference here is that at each step the function to be applied is chosen at random.

Let's consider this for the Sierpiński triangle with its template expressed by the three functions in (1). Take any initial starting point, the origin (0, 0) will do. Select one of the three functions at random, for example by tossing a die, and if it comes up 1 or 2 choose the first function, for 3 or 4 choose the second, and for

5 or 6 choose the third, and apply the chosen function to the initial point to get a second point. Throw the die again to select another function and apply it to this second point to get a third. Continuing in this way, we get a sequence of points, each obtained from the previous one by applying one of the functions chosen at random. These points will not be arbitrarily scattered across the picture but will jump around the Sierpiński triangle and after a time will fill out the whole fractal. Omitting the first 100 points (to allow the procedure to settle down) and plotting the next 10,000 points will give a good computer image of the Sierpiński triangle. This method is effective for computing a picture of the fractal defined by any family of contracting transformations. The chaos game was used to draw the fractals in Figure 12.

Self-affine fractals

An *affine transformation* is more general than a similarity, in that it scales by different ratios in different directions and so elongates objects, for example affine transformations transform squares into rectangles or parallelograms, and circles into ellipses. A *self-affine* fractal is one that is made up of smaller affine copies of itself. Self-affine fractals can be represented by templates, but now the smaller shapes are affine copies, rather than similar copies of the large shape. Typically the template will consist of a large square and smaller rectangles or even parallelograms. Figure 16 shows a self-affine fractal with its template with each of the three rectangles containing an affine copy of the whole. For self-affine fractals the component parts may be elongated in certain directions, and this allows a much greater range of fractals to be constructed than just self-similar ones. Figure 17 shows a self-affine fern and a self-affine tree, which on looking closely are each made up of 5 smaller affine copies—they each have templates consisting of just 5 rectangles or parallelograms (the templates have not been shown to avoid the pictures becoming too cluttered).

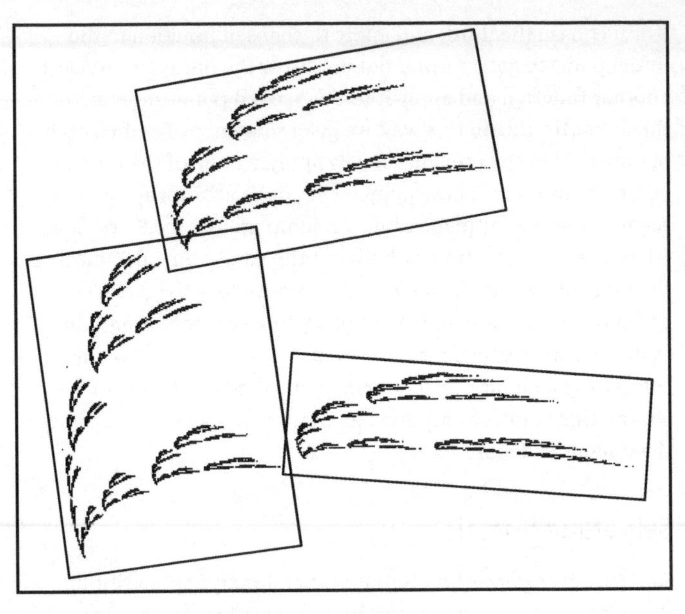

16. A self-affine fractal with its template

Statistically self-similar fractals

Another variation on self-similar constructions is to introduce some randomness. To illustrate this, let's return to the von Koch curve construction. As before, start with an interval, split it into three equal parts, and remove the centre part. But at this juncture toss a coin. If it comes up heads then insert the new line segments with the '∧' shape pointing upward, just as before, but if it is a tail insert the '∨' shape pointing downward rather than upwards. Now remove the middle thirds of these four straight segments, and toss a coin for each gap in turn. Insert the '∨' pointing upwards for each head, but downwards for each tail. Continue in this way, tossing the coin for each gap to determine the direction of the new part. This produces a *random von Koch curve*, see Figure 18.

17. A self-affine fern and tree

This random curve is not self-similar in the strict sense: 33⅓ per cent photo-reduction does not give exactly the same picture as each of the four component parts. Nevertheless, it is *statistically self-similar* in that the modification inside each segment follows

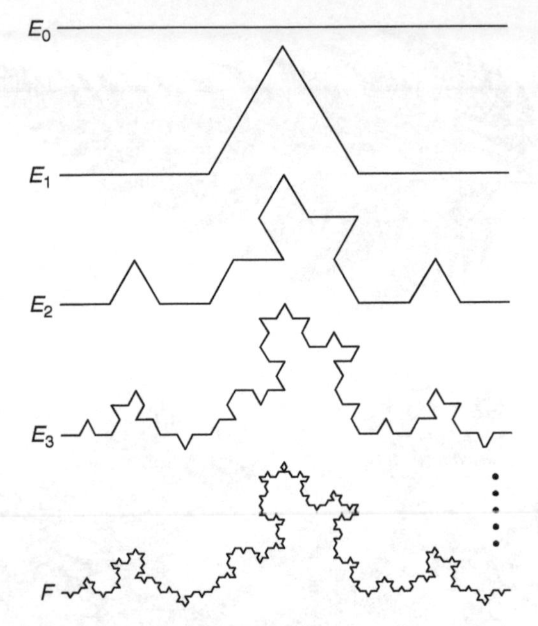

18. Construction of a random von Koch curve—when replacing each line segment a coin is tossed and the 'V' is drawn pointing upwards for a head and downwards for a tail

just the same random procedure at every stage of the construction.

Introducing an element of randomness into a fractal construction can often result in a more natural appearance. A 'real' coastline has irregularities more like the random von Koch curve than the non-random version. A forest skyline viewed from a distance might be considered a fractal, but there will be a randomness in the skyline resulting from statistical variations in tree heights.

Fractal image compression

Image compression involves coding pictures by a relatively small amount of information which is nevertheless enough to enable the accurate reconstruction of the pictures. Such techniques are central to internet technology. There is a limit to the rate at which data can be transmitted to a computer via a cable or a wireless connection and sending images pixel by pixel would be very slow indeed. To overcome this, pictures and movies are coded in a very efficient way before being sent to personal computers which contain software to decode the information and display the pictures.

Image compression depends on the fact that there is a considerable amount of redundancy in any picture or photograph. If the 2 million or so pixels or 'dots' on a computer screen were coloured completely haphazardly by any of the 16 million different shades in principle available, this information could not be compressed if an identical screen image was to be reproduced. However, in reality, regions of a picture will vary slowly in texture and colour, for example there may a large area of sky of nearly constant shade. Moreover, there may be repetition—one part of a wheat field looks very similar to another. A compression method capitalizes on such redundancy to reduce the amount of information needed to specify the picture. Such methods are lossy in that the picture reconstructed from the compressed data will not be identical to the original—pixels will have slightly different colours and some features may be less clear. Nevertheless, a good compression method will lead to a reconstructed image that, to the eye, is virtually indistinguishable from the original.

We have seen many fractals that are determined by a template of a simple form. In particular, convincing pictures of natural objects such as trees, clouds, ferns, or grasses result from templates with a small number of shapes. These pictures are completely specified by their templates (along with any orientations needed) and these can be represented very concisely, for example by giving the coordinates of the corners of the figures in the template. This provides a very efficient way of coding intricate pictures—the picture of the tree in Figure 17 is effectively determined by about 30 two-digit numbers. In this way complex objects can be specified by a small amount of data that allows the object to be reconstructed using one of the drawing methods indicated above.

As Michael Barnsley realized in the late 1980s, this has a potential for image compression. If, starting with some black and white drawing in the plane, one could find a relatively small number of transformations for which the drawing, or at least a good approximation to it, is the attractor, then the drawing could be compressed into a small amount of data. Ideally, one would like an automatic 'scanner' or 'camera' which could scan a picture and output a small family of transformations or, equivalently, a template, whose attractor is indistinguishable from the picture.

A great deal of research was done into the 1990s to develop such a process. One method, very roughly, involves partitioning a picture into squares and, for each 2×2 block of four squares, scanning the individual squares to find the one in which the picture is as closely as possible a $\frac{1}{2}$ scale version of that within the 2×2 block. This gives a similarity transformation from a block of squares to an individual square, and taking these all together yields a family of transformations with an attractor that is hopefully close to the original picture. Essentially, this procedure seeks out approximate self-similarities within the drawing. Such a method may be adapted to greyscale pictures, that is those made up of all shades from black to white, and indeed to colour pictures.

Fractal compression methods can give very good data compression ratios with little loss of resolution when the pictures are reconstructed. Moreover, reconstructing images from the family of transformations is very quick and efficient. The disadvantage is that the process of scanning squares required to encode images is computationally rather slow. In particular, the automatic encoding was not really rapid enough when the challenge became to compress videos in the mid- to late 1990s. Nowadays, methods such as JPEG and, for video, MPEG, together with wavelet methods dominate image compression and fractal compression is less used.

Chapter 3
Fractal dimension

A characteristic feature of fractals is their fine structure, that is their detail at arbitrarily small scales. 'Fractal dimension' attempts to quantify this by measuring the rate at which increased detail becomes apparent as we examine a fractal ever more closely. It indicates the complexity of the fractal and of the amount of space it occupies when viewed at high resolution. There are various definitions of dimension, but all depend on measuring fractals in some way at increasingly fine scales. In this chapter, we develop one of the most commonly used definitions, 'box-counting dimension', based on counting the number of squares in fine grids which overlap a fractal.

The inadequacy of length and area

Measuring, or perhaps calculating, the size of objects, is fundamental in mathematics, science, and, indeed, everyday life. But the way we measure something depends on its form. For a thin piece of wire or a road journey pencilled on a map we would measure a length. A wire is a 1-dimensional object, in that a single number, such as the distance along from one of its ends, is enough to specify any point along the wire. For a piece of card, a coloured region on a piece of paper, or a spherical surface, we might want to know the area, for example if we want to find how much paint is needed to cover the surface. These are

2-dimensional objects, with 2 coordinates needed to fix a position on them—in the case of a sphere both latitude and longitude are needed. The area of a rectangle is the product of its width and height and the area of other shapes can be estimated by dividing them into small rectangles (approximately if not exactly) and adding up their individual areas. Similarly for 3-dimensional objects, such as a solid cube or container of water, the volume is important. Length, area, and volume are the natural measures of 1-, 2-, and 3-dimensional objects. Using an inappropriate measure gives an unhelpful answer. Any line drawn with a fine pen on a piece of paper effectively has zero area since the ink covers negligible area. Similarly a thin piece of card has zero volume. On the other hand, it makes little sense to talk about the length of a piece of card—there is room on the card to draw a doodle whose length is as long as we wish. The card is too big to be measured by length or 1-dimensional measure, but too small to be measured by volume or 3-dimensional measure. The card is 2-dimensional and it is the area, the 2-dimensional measurement, that gives useful information about the card's size, with the card having positive, but not infinite, area.

In a similar manner, we seek a useful way of describing the 'size' of fractals. Let's return to the von Koch curve. We pointed out in Chapter 1 that if we measure the von Koch curve by taking very short steps around the curve and find the distance traversed, the answer grows without bound as the step length is reduced. To be a little more specific, we can work out the lengths of each stage of the construction of the von Koch curve in Figure 2. For convenience, assume that the initial line E_0 has length 1 unit. Then the first stage of the construction, E_1, comprises 4 straight pieces each of length 1/3, so has total length $4/3 = 1.333$. The next stage, E_2, is made up of $16 = 4^2$ pieces of length $1/9 = 1/3^2$, so has length $16/9 = 4^2/3^2 = (4/3)^2 = 1.778$ to three decimal places, where the superscript 2 indicates that the number is *squared* (for example, $4^2 = 4 \times 4$). Similarly, E_3 is made up of $64 = 4^3$ pieces of length $1/27 = 1/3^3$, so has length $64/27 = 4^3/3^3 = (4/3)^3 = 2.370$, where a superscript 3 means that the number is *cubed* (so $4^3 = 4 \times 4 \times 4$). Continuing in this way, the kth stage of the construction, E_k, consists of 4^k pieces each of length $1/3^k$, so the total length is $4^k/3^k = (4/3)^k = 1.333^k$, where the superscript k means that the number is *raised to the power* k, that is multiplied by itself k times (thus $4^k = 4 \times 4 \times \ldots \times 4$ with '4' occurring k times in this multiplication). These lengths of the stages of the construction increase very rapidly indeed, just like interest or inflation at 33.3 per cent! For example, E_5 has length $(4/3)^5 = 4.214$, E_{10} has length $(4/3)^{10} = 17.757$ and E_{50} has length $(4/3)^{50} = 1765780.963$. When k is very large, E_k includes many of the ins and outs of the von Koch curve and is a very close approximation to it. However, the closer the approximation the longer it is. The length of E_k is $(4/3)^k$ and this grows without bound as k gets large, so the von Koch curve has *infinite length*—it is too big to be measured in a useful way using length.

What about area? Just like a doodle on a piece of paper, if the von Koch curve is drawn with a very fine pen, even if it were possible to include every detail of the curve, the area of the page covered by the ink is negligible, so it has zero area.

Thus the von Koch curve has infinite length or 1-dimensional measure, and zero area or 2-dimensional measure—it is too big to be thought of as 1-dimensional and too small to be considered 2-dimensional. As we will see, it makes sense to regard the von Koch curve as having an intermediate dimension, in fact a 'fractional' dimension of about 1.262.

Box-counting dimension

There are a number of ways of defining the dimension of a fractal, but all involve measuring the fractal at various scales and seeing how these measurements behave as the scale becomes increasingly fine. Here we will use 'box-counting' to estimate how much of the plane the fractal fills when examined at small scales.

Given a set or shape, we superimpose a grid of squares or 'boxes' of small side lengths, and count the number of these boxes that overlap or (to use a mathematician's term) intersect the set. We do this using grids of a range of small sizes. To begin with, for a length 1 line, Figure 19(a) shows that 4 boxes of side $1/4$ overlap the segment, as do 8 boxes of side $1/8$. Continuing in this way, we can tabulate these numbers:

Line segment of length 1:

Side length of boxes	$1/2$	$1/4$	$1/8$	$1/16$	$1/32$	r
Number of boxes	$2 = 2^1$	$4 = 4^1$	$8 = 8^1$	$16 = 16^1$	$32 = 32^1$	$1/r = (1/r)^1$

where, in general, if we take a grid of side r the number of overlapping boxes will be the whole number immediately above $1/r$, and when r is small taking this number to be $1/r$ is close enough. If we do the same with a (solid) square of side 1, see Figure 19(b), $16 = 4^2$ boxes of side $1/4$ overlap the square, as do $64 = 8^2$ boxes of side $1/8$, and in general about $(1/r)^2$ boxes of side r:

Square of side 1:

Side length of boxes	$1/2$	$1/4$	$1/8$	$1/16$	$1/32$	r
Number of boxes	$4 = 2^2$	$16 = 4^2$	$64 = 8^2$	$256 = 16^2$	$1024 = 32^2$	$(1/r)^2$

The crucial thing to note from these two tables is that, when r is small, the number of boxes of side r that overlap the line segment

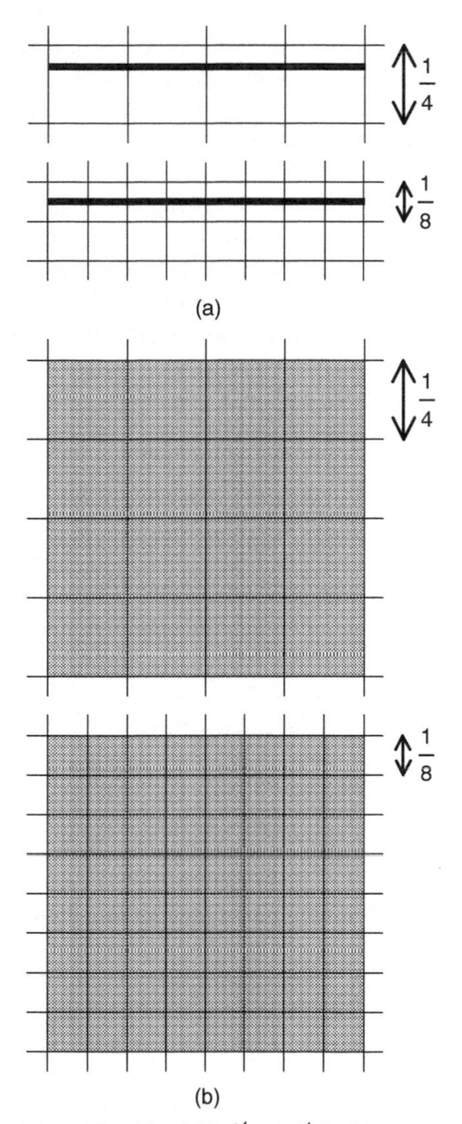

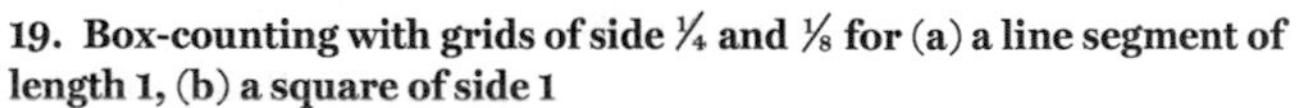

19. Box-counting with grids of side ¼ and ⅛ for (a) a line segment of length 1, (b) a square of side 1

is about $(1/r)^1$, whereas the number that overlap the square is $(1/r)^2$. The power 1 in the case of the line segment and the power 2 for the square are indicative of the fact that the segment and square have dimensions 1 and 2 respectively.

So let's try the same thing with the Sierpiński triangle (constructed inside a square of side 1). It is clear from the construction that 3 boxes of side ½ overlap the triangle. Figure 20 shows that 9 boxes of side ¼ and 27 boxes of side ⅛ overlap the triangle. Each time we halve the side length of the boxes the number of overlapping boxes is multiplied by 3, so continuing in this way, 3^k boxes of side $1/2^k$ overlap the Sierpiński triangle for each k.

By analogy with the line segment and square, we would like to express these box counts, 3, 9, 27, etc., as powers of 1/(sidelength), that is as powers of 2, 4, 8, etc. This can be done, except that the power required is no longer a whole number, in fact it is about 1.585:

Sierpiński triangle:

Side length of boxes	½	¼	⅛	1/16	r
Number of boxes	$3 = 2^{1.585}$	$9 = 4^{1.585}$	$27 = 8^{1.585}$	$81 = 16^{1.585}$	$\approx (1/r)^{1.585}$

Thus we consider the Sierpiński triangle to have dimension 1.585.

This raises the question of what is meant by raising a number to a fractional power, that is a power that is not a whole number. For whole number powers, this is just multiplying the number by itself that number of times, so

$$4^2 = 4 \times 4 = 16,\ 4^3 = 4 \times 4 \times 4 = 64,\ 4^4 = 4 \times 4 \times 4 \times 4 = 256,$$

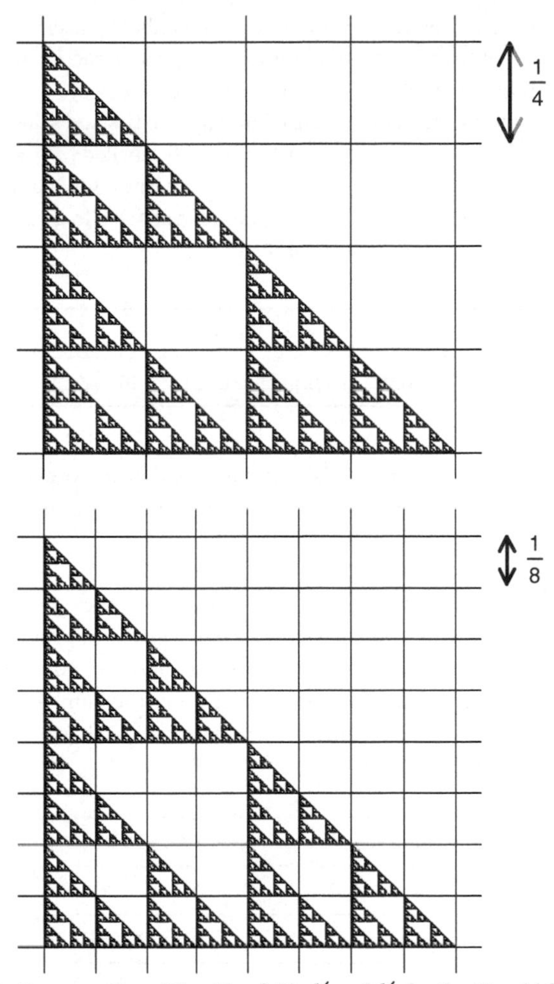

20. Box-counting with grids of side ¼ and ⅛ for the Sierpiński triangle

etc. But how can it make sense to multiply a number by itself 1.585 times? Suffice it to say that values may be assigned to fractional powers of positive numbers in a consistent and useful way. Fractional powers are frequently used in mathematics and science, and many calculators have a key for this, usually labelled [x^y], so, for example, the key sequence [4] [x^y] [1.4] returns 6.96, that is $4^{1.4}$, and [4] [x^y] [1.585] returns 9.0, the value of $4^{1.585}$. To give the idea, here are a few other powers of 4:

	4^1	$4^{1.2}$	$4^{1.4}$	$4^{1.6}$	$4^{1.8}$	4^2	$4^{2.2}$
=	4	5.28	6.96	9.19	12.13	16	21.11

Readers wishing to know more about the definition and properties of powers of numbers are referred to the Appendix at the end of this book.

Returning to box-counting, for brevity we call the grid made up of squares of side-length r the *grid of side r*. For a given shape in the plane, we write $N(r)$ as a shorthand for its *box-counting number* or *box-count* using boxes of side r, in other words the number of squares of the grid of side r that overlap the shape. We have seen that, for small r:

Line segment of length 1	$N(r) = (1/r)^1$
Square of side 1	$N(r) = (1/r)^2$
Sierpiński triangle	$N(r) = (1/r)^{1.585}$

with the powers 1, 2, and 1.585 giving the respective dimensions. We could do similar box-counts for lines, squares, or Sierpiński triangles of other sizes. For example, counting squares as before we get:

Line segment of length 2	$N(r) = 2(^1/_r)^1$
Line segment of length 3	$N(r) = 3(^1/_r)^1$
Square of side 2	$N(r) = 4(^1/_r)^2$
Square of side 3	$N(r) = 9(^1/_r)^2$
Sierpiński triangle of base length 2	$N(r) = 3(^1/_r)^{1.585}$

Again the power of $(^1/_r)$ indicates the dimension of the set, but this is multiplied by a number which, it may be noted, equals the length of the line segments, the area of the squares, and is somehow indicative of the overall size in the case of the Sierpiński triangle.

For line segments, squares, and the Sierpiński triangle, it was quite easy to do box-counts, at least for grids of side $^1/_2$, $^1/_4$, $^1/_8$, ... where the set fitted neatly into the grid. For other shapes it may be difficult to get exact numbers, but, none the less, careful counting or a little calculation allows good estimates of $N(r)$ to be found. For example, to a good approximation when r is small,

Circle of radius 1	$N(r) = 8(^1/_r)^1$
Circle of radius 3	$N(r) = 24(^1/_r)^1$
Von Koch curve	$N(r) = (^1/_r)^{1.26}$

indicating that the circles, as might be expected, have dimension 1, whereas the von Koch curve has dimension 1.26.

The formulae that we have just seen for the box-counting numbers all have a similar form. A formula such as $N(r) = 4(^1/_r)^2$ is called a *power law* because the number $^1/_r$ is raised to a fixed power—in this case 2. The power in question is called the *exponent* of the power law, and the number 4 that appears in front is called the *multiplier*, since it multiplies everything by a constant. So, for example, the box-counting numbers of the Sierpiński triangle of base length 2 satisfy a power law with exponent 1.585 and multiplier 3.

If the box-counting numbers for a fractal satisfy (exactly or approximately) a power law with exponent d, then, by analogy with the classical 1- and 2-dimensional shapes, we take d to be the dimension of the fractal. Thus a fractal has dimension 1.5 if its box-counting numbers are approximately $N(r) = 3(^1/_r)^{1.5}$, implying, for example, that the number of boxes of side 0.01 units that overlap the fractal is close to $3 \times (1/0.01)^{1.5} = 3 \times (100)^{1.5} = 3{,}000$.

The role of logarithms

Let's assume that the box-counting numbers of a fractal satisfy, at least approximately, a power law $N(r) = c\,(^1/_r)^d$ for some exponent d and some multiplier c. We can find box-counts $N(r)$ at various scales r, but how can we find the value of d from these counts? We need to be able to 'solve' this power law for d, and to do this we use logarithms.

Logarithms are closely related to powers of numbers. The *logarithm* of a number, abbreviated to log, is the power to which 10 must be raised to give that number, so log 100 = 2 since $10^2 = 100$ and log 1,000 = 3 since $10^3 = 1{,}000$. In general, the powers are not whole numbers, so as $10^{0.3010} = 2$ then log 2 = 0.3010. Here are a few more:

n	1	2	3	4	5	6	7	8	9	10
log n	0	0.3010	0.4771	0.6021	0.6990	0.7781	0.8451	0.9031	0.9542	1

Most calculators have a logarithm key, labelled [log], so, for example, the key sequence [log][5] returns 0.6990. Note that the larger a number, the larger is its logarithm.

A fundamental and very useful property of logarithms is that they turn multiplication into addition, that is for any positive numbers a and b,

$$\log(a \times b) = \log a + \log b. \tag{1}$$

If we multiply two numbers together, the answer is called the *product* of the numbers. Thus (1) is known as the *product rule* for logarithms. For example,

$$\log 12 = \log (3 \times 4) = \log 3 + \log 4 = 0.4771 + 0.6021 = 1.0792.$$

Until around 1970, before the days of pocket calculators, everyone learnt about logarithms in their first year at high school. Logarithm tables were universally used for multiplying together awkward numbers—one simply looked up the logarithms of the two numbers and added them to get the logarithm of the product, and then used tables again to convert back to the product itself.

An immediate consequence of the product rule is that, on putting $b = a$ in (1),

$$\log a^2 = \log (a \times a) = \log a + \log a = 2 \log a.$$

Thus the logarithm of the square of a number is just double the logarithm of the number. Similarly, the logarithm of the cube of a number is three times that of the number. In general, the logarithm of the dth power of a number is obtained by multiplying its logarithm by d, that is

$$\log a^d = d \log a;$$

this is the *power rule* for logarithms. For example, since $64 = 4^3$,

$$\log 64 = \log 4^3 = 3\log 4 = 3 \times 0.6021 = 1.8063.$$

More details about the product and power rules for logarithms are included in the Appendix.

Returning to the box-counting power law $N(r) = c\,(1/r)^d$, we want to find the dimension given by the exponent d. Taking logarithms of both sides and using first the product rule and then the power rule gives

$$\begin{aligned}\log N(r) &= \log(c\,(1/r)^d)\\ &= \log c + \log\,(1/r)^d\\ &= \log c + d\log\,(1/r).\end{aligned} \qquad (2)$$

Dividing everything by $\log(1/r)$ isolates d, so that

$$\frac{\log N(r)}{\log(1/r)} = \frac{\log c}{\log(1/r)} + d,$$

that is

$$d = \frac{\log N(r)}{\log(1/r)} - \frac{\log c}{\log(1/r)}.$$

At first sight, this seems unhelpful, since this expression still involves the multiplier c, which is not of particular interest when just trying to find the dimension d. However, in examining the fine structure of a fractal what matters is the box-counts when the grid side r is very small. And if r is small then $1/r$ and thus $\log(1/r)$ will be quite large, so $\log c / \log(1/r)$ will be small. Indeed, when r is very small, this term becomes negligible, so effectively $d = \log N(r)/\log(1/r)$. This remains true even if the power law is just a good approximation.

We conclude that $\log N(r)/\log(1/r)$ is close to the exponent in a box-counting power law when r is small. This motivates the formal definition of the dimension of a fractal, or indeed of any set, since it is equally valid for classical geometrical shapes.

The *dimension* of a fractal, sometimes called the *fractal dimension* or *box-counting dimension*, is defined to be

$$d = \lim \frac{\log N(r)}{\log (1/r)} \tag{3}$$

In this formula, $N(r)$ is the number of boxes of the grid of squares of size r that overlap the fractal, and 'lim', short for 'limit', is the number that $\log N(r)/\log(1/r)$ gets ever closer to as r gets very small.

To see how this works, let's try box-counting on the von Koch curve. Overlaying grids of various sizes on the curve and counting manually (which requires a little patience for the finer grids) we may obtain:

r	$1/4$	$1/8$	$1/16$	$1/32$	$1/64$	...
$N(r)$	6	14	33	78	189	...
$\log N(r)/\log(1/r)$	1.292	1.269	1.261	1.257	1.260	...

The bottom line shows $\log N(r)/\log(1/r)$ for each grid size. As r gets smaller and the grids become finer, this ratio gets near to 1.26. Indeed, $\log N(r)/\log(1/r)$ gets as close as desired to the number 1.2618... if r is taken small enough and using an extremely fine grid. We regard 1.2618... as the dimension of the von Koch curve, indicative that the box-counts satisfy a power law of the form $N(r) = c(1/2)^{1.2618...}$ for some multiplier c. In fact this dimension 1.2618... is just $\log 4 / \log 3$ which, as we will see later, is a consequence of the von Koch curve comprising 4 copies of itself at scale $1/3$.

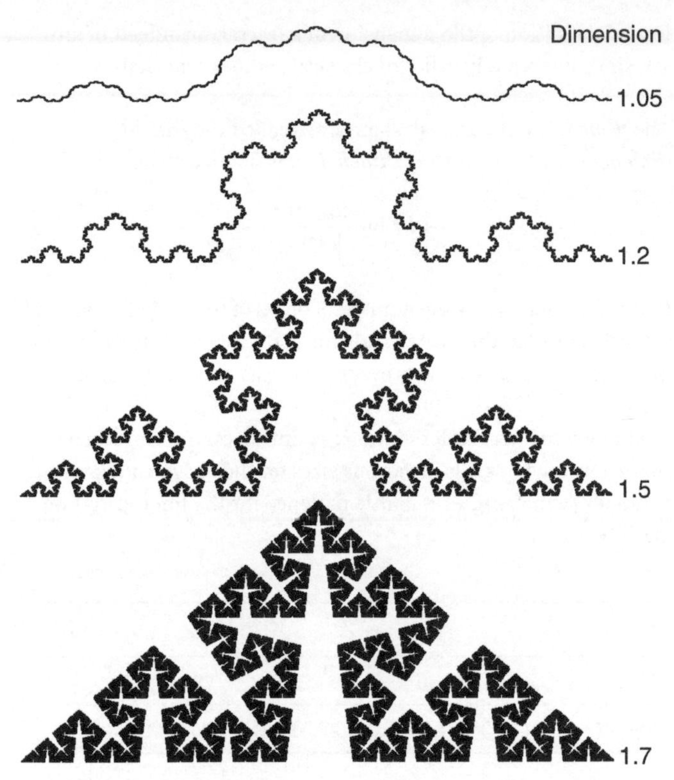

21. Variants on the von Koch curve with the dimensions depending on the angles used in the constructions

A similar approach enables the box counting dimension of many other fractals to be found. The snowflake and spiral in Figure 12 have dimensions about 1.535 and 1.585, respectively. Fractals in the plane can have any dimension between 0 and 2. By varying the angles in the motif used in the von Koch curve construction, all dimensions between 1 and 2 can be achieved, see Figure 21—as the irregularities of the curve becomes more pronounced the dimension increases.

Practical box-counting

Box-counting is a natural way to estimate the dimension of real fractals such as a fern, a forest skyline, a coastline, a crystal growth in a shallow dish, or the perimeter of a 'fluffy' cumulus cloud. Take a photograph of the fern, draw on it square grids of a range of sizes, and for each side length count the number of squares that overlap the fractal to get a list of numbers $N(r)$ and apply definition (3).

However, caution is needed. Mathematical fractals, such as the von Koch curve, are defined by a procedure that, in principle, goes on for ever, specifying the fractal in exact detail. They are fractal at all scales and the power law for the box-counts remains valid at arbitrarily fine scales. Real objects, however, only behave as fractals over a restricted range of scales. If the box sizes are too big then hardly any boxes will overlap the fractal and we get no useful information from a count. If the box sizes are too small then other factors dominate. A fern looks like the fractal in Figure 17 when viewed at scales between about 10cm and 5mm, but if inspected more closely solid stubby teeth are observed rather than scaled down replicas of the whole so the fractal assumption breaks down at fine scales. In reality, box-counts of real objects can only be expected to satisfy a power law over a limited range of scales at which the object displays a reasonably consistent degree of irregularity; we call this the *range of fractality*.

This poses a problem for estimating dimension. The von Koch curve is defined with complete precision at arbitrarily fine scales and we could, in principle, find box-counting numbers at scales as small as we like, either by photo-enlarging the curve and boxes to a size suitable for careful and tedious counting, or by doing some calculations. The definition of dimension (3) depends on the limit of what happens as the box sizes become infinitesimally small—the large scale box-counts don't matter. But for real fractals we cannot do this—we can only take box-counts over a limited range of scales and we have to harness this information to estimate the dimension as best we can.

Equation (2) expressed the power law $N(r) = c\,(1/r)^d$ in terms of logarithms as

$$\log N(r) = d \log(1/r) + \log c. \qquad (4)$$

If box-counts are available for very fine grids, so that r can be taken very small, then $\log N(r)$ and $\log(1/r)$ will be large and will dominate this equation, so $\log c$ can be neglected, leaving $\log N(r) / \log(1/r)$ as a fair estimate for d. However, if we only have box-counts $N(r)$ available for a range of relatively large box sizes when $\log(1/r)$ and $\log N(r)$ remain fairly small, then the term $\log c$ may be significant and we cannot ignore it.

Fortunately there's a way around this difficulty. If we draw the graph of the equation $y = 3x + 2$, so for each x we plot the point with coordinates $(x, 3x + 2)$, we get a straight line. The number 3 in this equation is the *slope* or *gradient* of this line, meaning that if we increase the value of x by a certain amount, the corresponding value of y given by the graph will increase by 3 times as much. More generally, for fixed numbers d and a, the graph of the equation $y = dx + a$ is a straight line with slope d; the slope does not depend at all on the number a. Thus if we know the (x, y)-values of a few points satisfying some straight line equation and draw a straight line through the points with these coordinates, the line will have slope d. This is the situation in equation (4). If, for the values of r at which we have taken box-counts, we plot the points with coordinates $(\log(1/r), \log N(r))$ on graph paper and draw the best straight line that we can through these points then, regardless of what $\log c$ is, the slope of this line will be d which we can take to be the dimension of the object regarded as a fractal over its range of fractality.

Let's look at the classic example of the coastline of Britain. The closer the coastline is examined, the more detail becomes apparent, so the coastline might be expected to behave like a fractal perhaps over a range of scales 100m–100km. By superimposing grids on suitable maps, see Figure 22(a), (showing squares of 60km side), we get the following box-counts:

Grid size r in km	140	100	60	30	20
Box count $N(r)$	33	55	103	217	399
$\log(1/r)$	−2.146	−2.0	−1.778	−1.477	−1.301
$\log N(r)$	1.519	1.74	2.012	2.336	2.601

The points with coordinates ($\log(1/r)$, $\log N(r)$) are plotted for each of these r on the graph in Figure 22(b) and the 'best' straight line, that passes as closely as possible to these points, is drawn (taking more account of the points corresponding to finer scales r where the box-counts provide more detailed information). The dimension is given by the slope of this line, which we measure by dividing a vertical increase v across part of the graph by the corresponding horizontal increase h. This is about 0.87/0.72 = 1.21, so we deem this to be the dimension of the coastline. Of course this is an approximation—the greater the range of scales over which box counts are available, the more meaningful the answer is likely to be. However, the fractality breaks down once the scales get sufficiently fine, indeed with variations of the water's edge and coastal geology, the exact line of the coast cannot even be meaningfully identified.

This procedure enables the dimension of many physical fractals to be found. For example, the dimension of a fern, estimated using boxes from 1 to 20cm is about 1.5, and the dimension of copper deposition by electrolysis in a flat dish (see Chapter 6) estimated on scales from 2mm to 5cm is about 1.7.

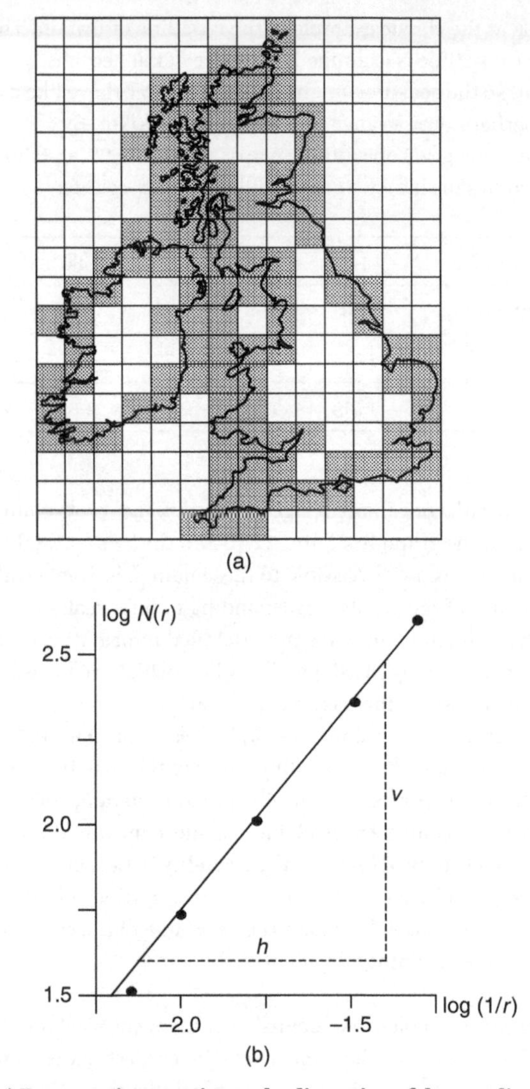

22. (a) Box-counting to estimate the dimension of the coastline of Britain, (b) a 'log-log' plot of the box counts—the slope v/h gives an estimate of the dimension

So far, our discussions have been restricted to fractals in the plane—this really is just for the convenience of drawing or visualizing them. In reality, objects belong to our three-dimensional world. There is no problem in principle in defining the dimension of fractals in space in the same way as in the plane. The dimension is given by the same formula (3) corresponding to the power that occurs in the power law behaviour of the box-counts, the difference is that now we superimpose a grid of *cubes* of side r and take $N(r)$ to be the number of these cubes that overlap the object. In practice, it is almost impossible to count the overlapping cubes. Dividing the space enclosing a plant, say, up into small cubes and counting how many cubes contain parts of the plant would be a considerable challenge! Nevertheless, there are ways around this which will be indicated shortly.

Dimension of self-similar fractals

To find the dimension of self-similar fractals we can utilize the self-similarity directly. When we found the dimension of the Sierpiński triangle by box-counting, we noted that each time we halved the side-length of the grid, the box-counting number was multiplied by 3, that is, for small r,

$$3N(r) = N(r/2),$$

see Figure 20. Assuming that the box-counting numbers satisfy a power law $N(r) = c(1/r)^d$, this means

$$\begin{aligned} 3c(1/r)^d &= c(1/(r/2))^d \\ &= c(2/r)^d \qquad (\text{since } 1/(r/2) = 2/r) \\ &= c\,2^d(1/r)^d. \end{aligned}$$

Cancelling $c(1/r)^d$ on each side of this equation leaves

$$3 = 2^d.$$

To find d we take logarithms of both sides and use the power rule, so

$$\log 3 = \log(2^d) = d \log 2,$$

and dividing by log 2 gives that $d = \log 3/\log 2$. The dimension of the Sierpiński triangle is the exponent in the power law for its box-counting numbers, so its dimension is $\log 3/\log 2$.

This little argument depends on the fact that the Sierpiński triangle comprises 3 similar copies of itself each scaled by a factor ½, and consequently that on halving the grid size the box-counting number is multiplied by 3. The dimension of other self-similar fractals may be found using a similar approach:

> A fractal that is made up of a similar copies of itself at scale $1/b$ has dimension $\log a/\log b$.

For instance, the von Koch curve consists of four copies of itself at scale ⅓, so has dimension $\log 4/\log 3 = 1.2618\ldots$, as before.

Measurement within dimension

Answering the (somewhat hackneyed) question 'How long is a piece of string?' with the words 'it's 1-dimensional', is of little help for most practical purposes. To know whether the string is any use for tying a parcel or hanging a picture, '50 centimetres' or '30 inches long' would be more helpful. When dealing with something that we know to be 1-dimensional, such as string, cotton, or wire, length provides further information on its size. Similarly, for a 2-dimensional object such as a sheet of paper or pane of glass, area is the appropriate measurement, whilst for a block of wood or glass of beer, volume, the 3-dimensional measure, is what is needed. There is a method of measurement appropriate for use within each of these dimensions—length, area, volume. Analogously, for each fractal dimension, there is a natural way of measuring objects within that dimension. To get the rough idea, let's return to the power laws—they contain information that we have so far ignored. Recall that the box-counts of the grid squares of side r that overlap solid squares of side 1, 2, and 3 are near enough $1(\frac{1}{r})^2$, $4(\frac{1}{r})^2$ and $9(\frac{1}{r})^2$ respectively when r is very small. The multipliers 1, 4, and 9 are recognizable as the areas of the squares of sides 1, 2, and 3. Thus if the box-counts for a square gives a power law of $c(\frac{1}{r})^2$, this not only says that it is 2-dimensional, but also suggests that its area, that is its 2-dimensional measurement, is given by the multiplier c. In fact it turns out that *any* shape in the plane satisfies a power law $c(\frac{1}{r})^2$ for small r, where the multiplier c is its area.

In the same way, if the box-counts for a fractal using fine grids satisfy a power law $N(r) = c(\frac{1}{r})^d$ for some c and d, not only does this tell us that the dimension of the fractal is d but also indicates that it is of 'size' c within the collection of d-dimensional fractals. We can think of c as a 'd-dimensional measurement' of the fractal, in the sense that area is a 2-dimensional measurement.

Now consider the way measurements change under scaling. If we scale by a factor 2 (that is, do a 200 per cent enlargement on a photocopier) the lengths of lines and curves are multiplied by 2, but the area of squares, circles, etc., are multiplied by $2^2 = 4$, see Figure 23. Similarly scaling by a factor 3 (or 300 per cent) multiplies lengths by 3 and areas by $3^2 = 9$. Our d-dimensional measurement scales in the same sort of way, but the power involved is just the dimension d. Recall that the box-count for the von Koch curve satisfies an (approximate) power law $N(r) = (\frac{1}{r})^{1.26}$. So if we consider the square grid of side $\frac{r}{2}$ (as opposed to r), about $(\frac{2}{r})^{1.26}$ of the boxes overlap the curve. Enlarging this whole picture by a factor 2 enlarges the grid boxes to side length r, of which the same number $(\frac{2}{r})^{1.26} = 2^{1.26}(\frac{1}{r})^{1.26}$ overlap the enlarged fractal. Thus the enlarged fractal has box-counting numbers $N(r) = 2^{1.26}(\frac{1}{r})^{1.26}$. The multiplier in the power law is now $2^{1.26}$ so scaling the von Koch curve by a factor 2 has multiplied its 1.26-dimensional

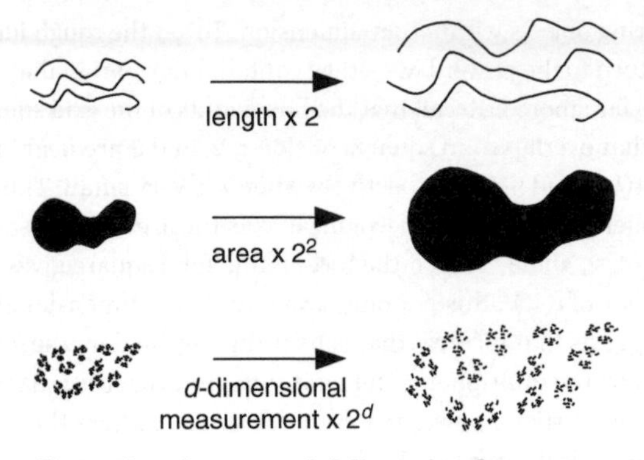

23. The effect on length, area, and *d*-dimensional measurement of enlargement by a factor 2

measurement by $2^{1.26} = 2.39$. Similarly, scaling by a factor 3 multiplies the measurement by $3^{1.26} = 3.99$, and so on. The same holds for any fractal of any dimension d: scaling by 2 multiplies the d-dimensional measurement by 2^d, scaling by 3 multiplies it by 3^d, etc.

In fact, there are difficulties in taking the multiplier c in a power law as a 'd-dimensional measurement'; to begin with it assumes that there is actually a constant multiplier, whereas for many fractals it 'wobbles about a bit'. However, this heuristic argument is a first stab at generalizing the familiar concepts of length, area, and volume to fractional dimension in a natural way. This idea leads to what is known as d-dimensional Hausdorff measure, named after the German mathematician Felix Hausdorff (1868–1942), which is widely used by mathematicians.

Properties of dimension

For dimension to be a worthwhile concept, it should have properties that are both natural and useful. The first thing that box-counting dimension achieves is to extend the classical notion of dimension to a much wider setting. The box dimension of smooth curves such as circles or ellipses is 1 and of surfaces or plane regions such as solid squares is 2, consistent with the traditional concept of dimension as a whole number. However, box-dimension applies to essentially any geometric objects, not just curves and surfaces. This approach of taking a familiar idea and finding a way to generalize it to a wider setting is regularly seen in mathematics and has led to many significant developments.

Dimension has several properties without which it would be of little use. Firstly, 'the larger an object, the larger its dimension'. For example, erasing any part of the Sierpiński triangle gives a smaller set whose dimension therefore can be no larger than 1.585, the dimension of the original Sierpiński triangle.

If a set comprises just a finite number of individual points scattered around the plane, its dimension is 0. On the other hand the dimension of a set in the plane cannot be more than 2, the dimension of the plane itself. Similarly, if we consider fractals in 3-dimensional space, where box-counting dimension involves counting cubes, then dimensions can be anything up to 3, the dimension of space itself. (Strange spaces are encountered in some areas of mathematics or physics where dimensions can be larger, but these need not concern us here.)

A continuous curve in the plane always has dimension at least 1. Conversely, if a fractal has dimension less than 1 then it must be 'dust-like' or 'totally disconnected', that is, the fractal is so sparse that it is impossible to travel between any two points without going outside the fractal. For if it were possible to trace a continuous path inside the fractal then the fractal's dimension would be at least that of the path, that is at least 1.

Given two fractals in the plane, for example two fractal ferns, we might wish to lump them together and regard them as a single object, with its dimension given by box-counting across the entire picture. Then the dimension of the combined object is (normally) the larger of the two individual dimensions—the fractal with the larger dimension 'dominates' the other. Thus a von Koch curve with a straight line adjoined to one end has dimension 1.26, the larger of 1.26 and 1.

Dimension is a stable quantity—distorting a fractal in a smooth manner does not change its dimension. If a fractal is drawn on a thin sheet of rubber which is then stretched or deformed smoothly, without introducing any folds, corners, or kinks, its box-counting dimension is unchanged. In particular transforming a fractal by a similarity does not affect the dimension (though it will change any d-dimensional measurement).

We might look at the *section* of a fractal by a straight line, that is the set of points of the fractal that lie on the line. Dimensions of sections are generally well-behaved. As an illustration, draw a straight line through the Sierpiński triangle, Figure 11, in an arbitrary direction. The section will be a dust-like fractal consisting of those points scattered along the line that are in the Sierpiński triangle. The dimension of this fractal will almost certainly be 0.585 = 1.585 – 1, that is the dimension of the Sierpiński triangle minus 1. Of course, as can be seen from the picture, for some lines the overlap with the Sierpiński triangle will be an entire line segment and therefore of dimension 1, for example in certain positions when the line is horizontal. However, such positions are exceptional, and if the line is drawn 'at random' a section of dimension 0.585 is likely. In general, 'typical' or 'random' sections of a fractal by a straight line have a dimension that is 1 less than that of the fractal, unless this number is negative, in which case the fractal is so sparse that it will have no points in common with 'most' lines. Looking at part of the fractal within a 1-dimensional line rather than at the whole within 2-dimensional space results in this dimension decrease of 1.

Dimensions also behave well under *projections* 'or shadows'. Here it is simplest to think of a fractal in 3-dimensional space, a plant perhaps. Looking at its shadow or taking a photograph gives a flat picture of the plant (a 'projection onto a plane') whose dimension might be found by box-counting. The dimensions of a typical photograph and the original plant are related: if the dimension of the plant is greater than 2 then its picture will have dimension 2 and have positive area; if the dimension of the plant is less than 2 then the dimension of the picture will be the same as that of the plant. (This is a rather vague paraphrase of a precise and important mathematical theorem about projections.) This principle enables the dimension of some spatial objects to be estimated in a practicable way by box-counting on a photograph.

Limitations of dimension

Whilst dimension provides fundamental information about the character of a fractal and its appearance under ever-increasing magnification, a single number can still only provide limited information about objects that come in a enormous variety of shapes and forms.

Fractals of a given dimension may have widely differing structures. For example, a fractal of dimension 1.26 could be a continuous curve such as the von Koch curve, it might be full of holes rather like the Sierpiński triangle, it might be a 'dendrite' or 'twig-like', it could be totally disconnected or 'dust-like', or it may be 'striated', consisting of irregularly spaced parallel line segments. There is a major branch of mathematics known as *topology* concerned with distinctions of this nature.

Dimension tells us little about the 'texture' of fractals, and various quantities have been introduced to describe such features. These include lacunarity (from the Latin *lacuna* meaning hole) and porosity (from the Greek *poros* meaning passage) which indicate the size distribution of holes in an object, for example, the

Sierpiński triangle has high lacunarity. Such concepts are important in applications such as geology and soil science where a crucial factor is the structure of the network of gaps between the solid parts through which oil or nitrates can flow.

Other notions have been proposed to describe different aspects of 'fractality' but these often have rather technical definitions and can be awkward to work with both in theory and in practice. For example, one can look at the behaviour of 'local dimensions', that is the way the dimension of the part of a fractal within a small disc varies as the disc is moved around. For real fractals, however, the range of fractality is rarely wide enough to allow meaningful measurements of such quantities.

Despite its limitations, dimension stands out as a very useful and robust quantity that codifies the way objects scale, something that is important in many contexts. Dimensions are defined in a fairly straightforward way and are relatively easy to measure in practice, by box-counting or other techniques, as well as being theoretically tractable. It is not surprising that dimension has become a major tool in studying the range of fractals that are encountered across mathematics, science, and social science, but caution is nevertheless needed in its use and interpretation.

Chapter 4
Julia sets and the Mandelbrot set

Julia sets and the Mandelbrot set are amongst the most frequently pictured fractals, combining both aesthetic and actual beauty. They have been used as a basis for modern art, as scenery in science fiction films, and simply as mystical symbols. The complexity of the Mandelbrot set is mind-boggling and its mathematical properties are far from fully understood, yet its definition is simple and it can be realized on a computer screen with just a few lines of computer code. Like some of the other fractals we have encountered it is determined by the behaviour of itineraries around the plane—once again repeating a simple step over and over again leads to extraordinarily complicated objects.

It is possible to think of Julia sets and the Mandelbrot set purely in terms of iteration of functions expressed in coordinate form. However, a little knowledge of complex numbers highlights the simple elegance of these functions.

Complex numbers

By definition, the *square root* of a number is a number which, when squared, gives that number. For example, since $3^2 = (-3)^2 = 9$, the square roots of 9 are 3 and -3. A quirk of the usual or *real* number system is that, whilst positive numbers have square roots, negative numbers do not. As the square of every number is non-negative there is no real number whose square is –9. Hence –9, –100, –¼, and all other negative numbers do not have a square root. This asymmetrical situation unsettled mathematicians for thousands of years and it was not until the 16th century that the matter was resolved by inventing a new entity, namely the square root of the simplest negative number, –1.

Thus, we bring in to the usual number system a 'number' which we think of as $\sqrt{-1}$, the square root of –1. It is denoted by i and has the defining property that its square is –1, so $i^2 = i \times i = -1$. This enables us to express the square roots of *all* negative numbers in terms of i. For example, $\sqrt{-9} = 3i$, since $(3i)^2 = 3^2 \times i^2 = 9 \times (-1) = -9$. Similarly $\sqrt{-100} = 10i$ and $\sqrt{-1/4} = \frac{1}{2}i$. This leads to an enlarged number system in which we can add, subtract, multiply, and divide in a consistent way.

A 'number' of the form $x + yi$ is called a *complex number* where x and y are ordinary real numbers and i is thought of as the square root of –1. We call x and y the *real part* and the *imaginary part* of $x + yi$ respectively. For example,

$$1 + 3i,\ 2 + i,\ 3 - 5i,\ 0 - 3i,\ \tfrac{1}{2} + \tfrac{3}{4}i,\ -1.5 + 2.8i$$

are complex numbers, and the number 2 + 3i has real part 2 and imaginary part 3. It is often convenient to think of complex numbers as a single entity, so we may write $z = x + yi$ and refer to this as 'the complex number z'.

The basic arithmetic operations of addition, subtraction, and multiplication of complex numbers work in a natural way so that all the usual rules of arithmetic and algebra hold, except that we replace i^2 by –1 whenever it is encountered, consistent with i being the square root of –1. Here are some examples:

$$(1 + 3i) + (2 + i) = (1 + 2) + (3i + i) = 3 + 4i$$
$$(1 + 3i) - (2 + i) = (1 - 2) + (3i - i) = -1 + 2i.$$

To multiply two complex numbers we multiply each part of the first number by each part of the second number, in a similar way to multiplying out an algebraic expression. For instance, here's how we square the complex number $2 + i$:

$$\begin{aligned}(2 + i) \times (2 + i) &= (2 \times 2) + (2 \times i) + (i \times 2) + (i \times i) \\ &= 4 + 2i + 2i + i^2 \\ &= 4 + 4i - 1 \\ &= 3 + 4i,\end{aligned}$$

where we have used that $i^2 = -1$. Thus $(2 + i)^2 = 3 + 4i$.

Complex numbers have a natural pictorial or geometrical representation. We simply think of the complex number $z = x + yi$ as located at the point with coordinates (x, y) in the Euclidean plane, which in this context is referred to as the *complex plane*. Thus we may regard points in the plane in two ways: either given by coordinates or as complex numbers. The point with coordinates $(4, 2)$ can equally well be thought of as the complex number $4 + 2i$. The locations of various complex numbers are shown in Figure 24. Note that the origin is the complex number $0 = 0 + 0i$.

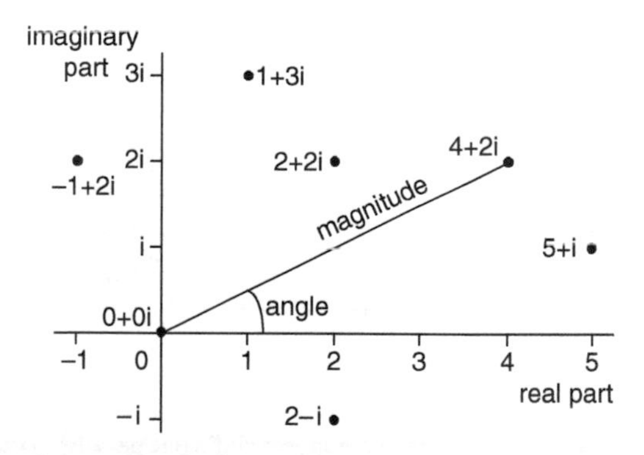

24. Complex numbers in the complex plane, with the magnitude and angle of 4 + 2i indicated

There is another convenient way of specifying the position of a complex number z in the plane. We simply draw a line from the origin to z. The length of this line, in other words the distance of z from the origin 0, is called the *magnitude* (or *modulus*) of z, and the angle between this line and the horizontal axis (measured in a counter-clockwise direction) is called the *angle* (or *argument*) of z, see Figure 24. If we know both the angle and magnitude of a complex number z, we know in which direction and how far away it is from the origin, and this is enough to determine its location.

Arithmetic operations on complex numbers have geometric interpretations in the complex plane. One way to view addition of two complex numbers, for example $(1 + 3i) + (2 + i)$, is to start at $1 + 3i$ and add 2 to the real part and 1 to the imaginary part, which corresponds to a shift from $1 + 3i$ parallel to and of length equal to the line joining 0 to $2 + i$, see Figure 25(a). Equivalently, the sum $3 + 4i$ is located at the fourth vertex of the parallelogram whose other three vertices are 0, $2 + i$ and $1 + 3i$. In general,

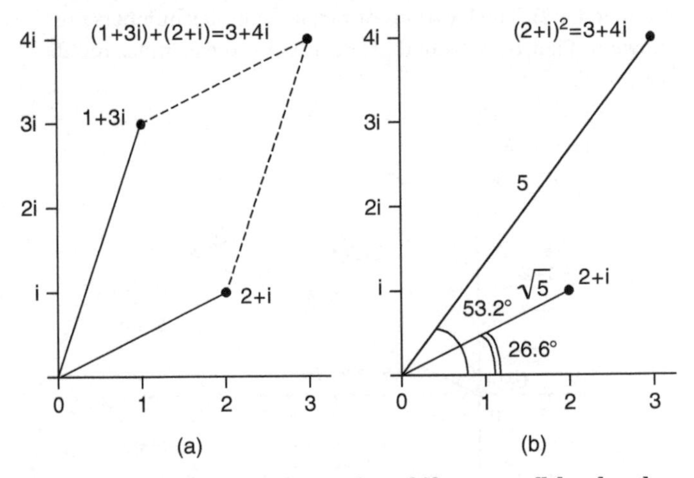

25. (a) Addition of two complex numbers shifts one parallel to the other, (b) squaring a complex number squares the magnitude and doubles the angle

addition of any two complex numbers may be thought of geometrically by shifting one of them in a direction and distance given by the line repesenting the other.

Pythagoras' Theorem provides a straightforward way to find the magnitude of complex numbers. To get from the origin to the complex number 2 + i (the point with coordinates (2, 1)) we step a distance 2 to the right and then step a distance 1 up. But the line joining the origin to 2 + i is the hypotenuse of the right-angled triangle with these two steps as the other sides. Pythagoras' Theorem tells us that the square of the length of this hypotenuse is $2^2 + 1^2$, so that the magnitude of 2 + i is $\sqrt{2^2+1^2} = \sqrt{5}$. In just the same way, the magnitude of a general complex number $z = x + y$i is $\sqrt{x^2+y^2}$.

We saw above that $(2 + \mathrm{i})^2 = 3 + 4\mathrm{i}$. The magnitude of 3 + 4i is $\sqrt{3^2+4^2} = \sqrt{25} = 5$ which is thc square of $\sqrt{5}$, the magnitude of 2 + i. We can also calculate the angles of these complex numbers: the angle of 3 + 4i is $53.130°$ which is exactly double $26.565°$, the angle of 2 + i, see Figure 25(b). Squaring 2 + i gives a complex number of double the angle and the square of the magnitude. This is no coincidence, but is an instance of a remarkable property of complex numbers:

> Squaring a complex number squares the magnitude and doubles the angle

Put another way, for every complex number z, the magnitude of z^2 is the square of that of z, and the angle of z^2 is double that of z. Verifying this property is a little involved, and more details are given in the Appendix.

Throughout this chapter we will be interested in a combination of these two operations, squaring and addition. We will be particularly concerned with the function

$$z \to z^2 + c \qquad (1)$$

which takes a complex number z, squares it, and adds a fixed complex number c, moving each point z in the complex plane to the point $z^2 + c$. From the preceding remarks the geometric effect of (1) is to square the magnitude and double the angle of z and then shift by the complex number c.

For any given c the effect of the function (1) on any complex number is easily calculated, for example, taking $c = 2 + \mathrm{i}$,

$$\begin{aligned} 3 + 4\mathrm{i} \to & (3 + 4\mathrm{i})^2 + (2 + \mathrm{i}) \\ & = (-7 + 24\mathrm{i}) + (2+\mathrm{i}) \\ & = -5 + 25\mathrm{i}, \end{aligned}$$

and similarly

$$1 + 2\mathrm{i} \to -1 + 5\mathrm{i} \text{ and } 5 - \mathrm{i} \to 26 - 9\mathrm{i}.$$

Written as $z \to z^2 + c$ the function (1) has a simple and pleasing form. For purposes of calculation and computation, it is useful to have a coordinate version of the function. Expressing the complex numbers in terms of their real and imaginary parts, so $z = x + y\mathrm{i}$ and $c = a + b\mathrm{i}$, (1) becomes

$$x + y\mathrm{i} \to (x + y\mathrm{i})^2 + (a + b\mathrm{i})$$

which, after squaring out the first term (see the Appendix) and adding, reduces to

$$x + y\mathrm{i} \to (x^2 - y^2 + a) + (2\,xy + b)\mathrm{i}.$$

Regarding complex numbers in terms of coordinates given by their real and imaginary parts, we may read off (1) in coordinate form as

$$(x, y) \to (x^2 - y^2 + a, 2\,xy + b). \qquad (2)$$

Iteration and Julia sets

A crucial thing to bear in mind for the remainder of this chapter is that a function is just a rule that tells us how to move around the plane—from each point in the plane the function takes us to some other point. As we saw in Chapter 1, starting at some initial point, repeated application or iteration of a function (which was expressed in terms of coordinates) takes us on an itinerary or tour around the plane. For a given function, the itinerary, and in particular where we get to in the long run, can depend on the starting point. We now examine the itineraries of functions given in terms of complex numbers by

$$z \to z^2 + c, \qquad (3)$$

as we have pointed out, this is equivalent to iterating the coordinate form (2).

We start with the simplest case where the constant c is 0, that is

$$z \to z^2 \qquad (4)$$

which just squares complex numbers; geometrically it squares their magnitudes and doubles their angles. Let's look at the itineraries of various initial complex numbers under this function. Starting at $1 + 2i$ the function (4) takes us to $(1 + 2i)^2 = -3 + 4i$. The next iterate is $(-3 + 4i)^2 = -7 - 24i$, and so on. The first few points of the itinerary starting at $1 + 2i$ are

$$1 + 2i \to -3 + 4i \to -7 - 24i \to -527 + 336i \to 164833 - 354144i \to \dots$$

These complex numbers get large very rapidly: starting at $1 + 2i$ the iterates rapidly rush off the page into the far distance; we say that they 'go off to infinity'. Indeed, since squaring a complex number squares its magnitude, the magnitude is squared at each step. The magnitude of $1 + 2i$ is $\sqrt{5}$ so the points on this itinerary have magnitudes $\sqrt{5}$, 5, 25, 625, 390625, ...

Now suppose that we start nearer the origin, say at $0.5 + 0.3i$ which has magnitude 0.583, and repeatedly apply (4). This time the itinerary is

$$0.5 + 0.3i \to 0.16 + 0.3i \to -0.0644 + 0.096i \to -0.00507 - 0.01236i \\ \to -0.00013 + 0.00013i \to \ldots$$

Again the magnitude is squared at each step, giving magnitudes of 0.583, 0.34, 0.116, 0.0134, 0.00018,...; squaring a number less than 1 yields a much smaller number so the itinerary rapidly approaches $0 = 0 + 0i$.

These two examples are typical of the itineraries of the function (4). Depending on the starting point, either the iterates move rapidly towards the origin, or they shoot off into the distance, never to return. (With the children's treasure hunt analogy of iteration, either a child rapidly comes back home, or wanders further and further away, never to be seen again!)

We can be more specific by considering the magnitude of the iterates, recalling that squaring a complex number squares its magnitude. Squaring a number greater than 1 gives a rather larger number, so if the magnitude of the initial point is bigger than 1 the magnitudes of the iterates rapidly get large on repeated squaring. On the other hand, squaring a number smaller than 1 gives an even smaller number, so iteration from an initial number of magnitude less than 1 gives an itinerary that approaches the origin. Geometrically, if a circle is drawn in the plane with centre at the origin and with radius 1, an itinerary starting at any point inside the circle will approach the origin, but an itinerary starting anywhere outside the circle will shoot off into the distance, Figure 26(a). The circle with centre the origin and radius 1 plays a special role. It is

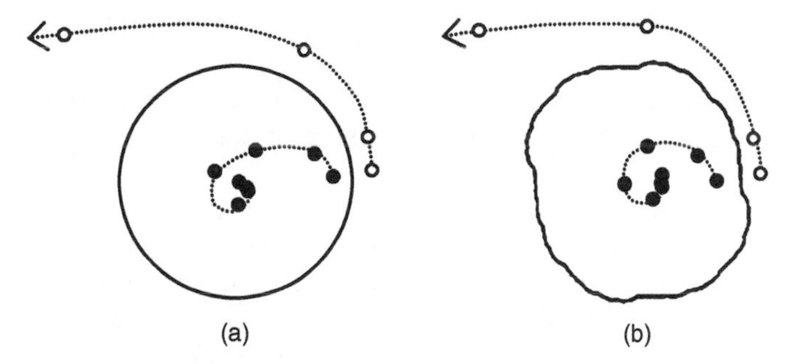

26. Itineraries under (a) $z \to z^2$ and (b) $z \to z^2 + 0.1 + 0.1i$, showing the curves that separate the two types of itineraries

the boundary between these two dramatically different types of behaviour, that is the boundary between the set of initial points with itineraries that approach the origin and those with itineraries that go to infinity. This boundary circle is termed the *Julia set* of the function (4). If we start with a point on the circle itself then the iterates will always remain on the circle, though this is an unstable situation as near any point on the circle there are points just inside that iterate to the origin and others just outside that iterate to infinity. This is an example of 'sensitive dependence on initial conditions'—a very slight change of the initial point can result in completely different long term behaviour of the itineraries.

Let's now change the function (4) slightly by adding some small complex number c, say $c = 0.1 + 0.05i$, so (3) becomes

$$z \to z^2 + 0.1 + 0.1i. \qquad (5)$$

This slight modification to the function might be expected to result in only a slight change in the behaviour of the iterates. Here are two itineraries under (5) with different starting points:

$$2 + i \to 3.1 + 4.1i \to -7.1 + 25.52i \to -600.76 - 362.28i$$
$$\to 229663.46 + 435291.86i \to \ldots$$

and

$$0.5 + 0.3i \to 0.16 + 0.4i \to 0.0076 + 0.308i \to 0.0052 + 0.1047i \to 0.0891 + 0.1010i \to 0.0977 + 0.118i \to \ldots$$

Starting at 2 + i the coordinates again get very large very rapidly, with the iterates rushing into the distance. On the other hand, starting at 0.5 + 0.3i the itinerary remains quite close to the origin. Geometrically, at each step we square the modulus and double the angle of the number and then shift by the complex number $c = 0.1 + 0.1i$; since c is small the squaring part dominates the shift so the behaviour is not very different from the case when $c = 0$. Just as with the function (4) two types of behaviour are possible depending on the starting point: either the iterates rapidly go off to infinity, or they stay near the origin. (In fact, in the latter case, the iterates always approach a particular point which is, to three decimal places, 0.094 + 0.123i. This is a *fixed point* of the function (5), in that it is unmoved by the function, so 0.094 + 0.123i → 0.094 + 0.123i. Starting the iteration at 0.094 + 0.123i the itinerary remains there for ever.)

The initial points that give the two types of behaviour under iteration by (5) are again separated by a curve. The picture is a just distorted version of that for (4): the separating curve is still a closed loop or circuit, but is no longer quite circular, see Figure 26(b). Indeed, on close inspection the curve appears very wiggly—it is in fact a fractal with dimension about 1.01. Repeated application of the simple function (5) leads to a fractal—the curve that separates the initial points of itineraries which go off to infinity from those initial points whose itineraries remain near the origin.

This motivates an important definition which, in principle, applies to any function. The set of initial complex numbers, thought of as points in the plane, from which the itinerary does *not* wander off to infinity is called the *filled-in Julia set* and the boundary of the filled-in Julia set is called the *Julia* set of the function. The Julia set separates two very different types of behaviour. If a point is in the Julia set there are arbitrarily close points that iterate to infinity but also arbitrarily close points that do not wander far off under iteration. For the function (4) the filled-in Julia set is the disc with centre the origin and radius 1 and its bounding circle is the Julia set. This picture is distorted a little for function (5), with the Julia set a fractal loop with the filled-in Julia set the region inside the loop.

If iterates of a complex number go off to infinity, then, unless the number lies extremely close to the Julia set itself, the iterates move off into the distance very rapidly indeed. This provides a way of computing the Julia set of a function. To see if a point is in the filled-in Julia set we examine its iterates under the function. If some iterate has magnitude more than 5, that is if the itinerary leaves the circle with centre the origin and radius 5, then the starting point is not in the filled-in Julia set. However, if the first 100 iterates all have magnitude less than 5 and remain in this

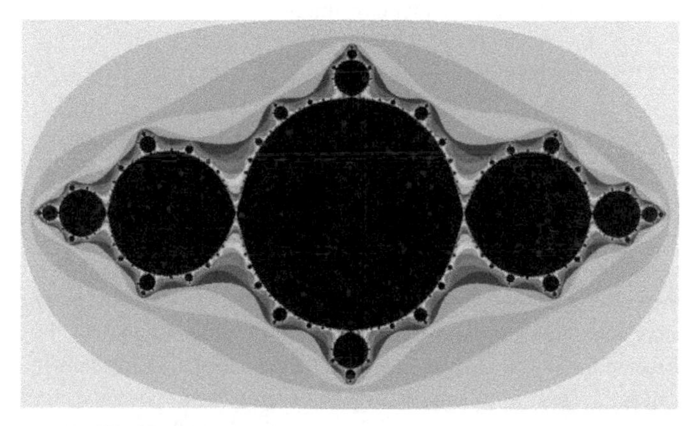

27. The filled-in Julia set of $z \to z^2 - 0.9$ with its exterior shaded according to the escape time

circle then the starting point is deemed to be in the filled-in Julia set and coloured (usually) black. This procedure is done for a large number of initial points in the plane, perhaps for each pixel on a computer screen. The black region of the plane obtained in this way is the filled-in Julia set, and its boundary is the Julia set. Often the points outside the filled-in Julia set are coloured or shaded according to the 'escape time', that is the number of iterations needed to reach distance 5 from the origin. The filled-in Julia set of Figure 27 was produced in this way. (The numbers 5 and 100 indicated here are typical values, but in practice are chosen depending on the function and on the accuracy of the picture required. From theoretical considerations, if some iterate of a point under (3) has magnitude that is greater than both the magnitude of c and 2, then the itinerary must go to infinity, a fact that reduces computation considerably. Moreover, apart from those pixels very close to the Julia set, there are techniques for considering large blocks of pixels together when testing the iterates.)

The zoo of Julia sets

So what happens if we iterate our function

$$z \to z^2 + c$$

for other complex numbers $c = a + b\mathrm{i}$? (As noted before, this may be written in terms of coordinates as $(x, y) \to (x^2 - y^2 + a, 2\,xy + b)$.) Geometrically, this function squares the magnitude and doubles the angle of z and shifts by c. We have seen that when $c = 0$ the Julia set is a circle, and when $c = 0.1 + 0.05\mathrm{i}$ it is a fractal loop or circuit. Is the Julia set always a loop? For complex numbers $c = a + b\mathrm{i}$ of fairly small magnitude, including all c with magnitude less than 0.25, the effect of the shift by c is not enough for the picture to change much, and the Julia set is still a loop, though its irregularities become more prominent as the magnitude of c increases, reflecting a larger dimension, see Figure 28(a),(b). However, if the magnitude of c gets

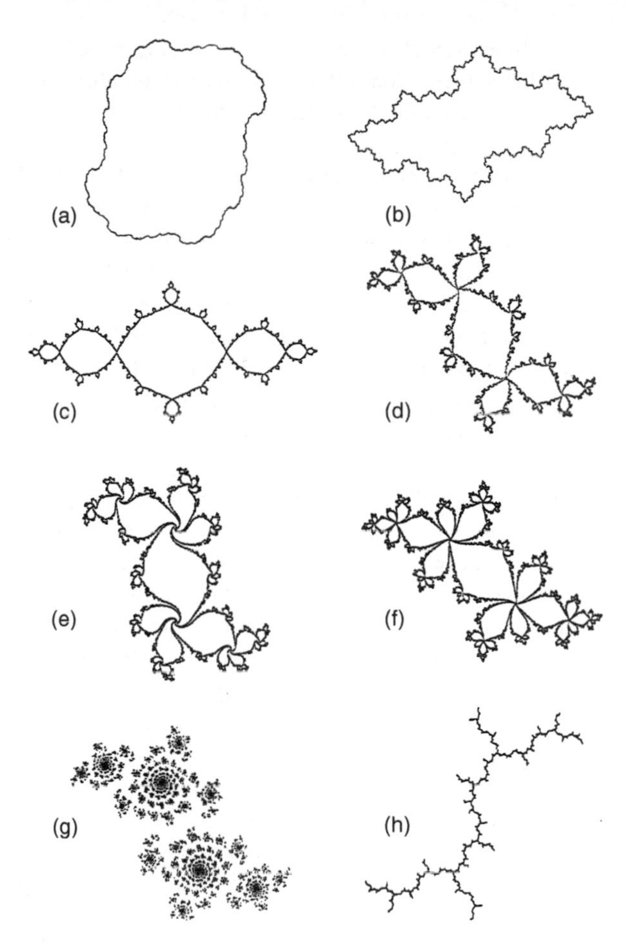

28. Julia sets for a range of complex numbers *c*, given by (a) 0.2–0.2i, (b) –0.6–0.3i, (c) –1, (d) –0.1 + 0.75i, (e) 0.25 + 0.52i, (f) –0.5 + 0.55i, (g) 0.66i, (h) –i

larger still and the shifting part of the function becomes more significant, strange things may happen. For example, with $c = -1$, the Julia set of the function $z \to z^2 - 1$ consists of a central fractal loop surrounded by many smaller touching loops which in turn are surrounded by even smaller loops, and so on, see Figure 28(c). Moreover, this configuration is stable: for complex numbers c close to -1 the Julia set has a similar structure of touching loops.

Investigating Julia sets of the function $z \rightarrow z^2 + c$ for other complex numbers c yields more surprises. With c close to $-0.1 + 0.75i$ the Julia set is again formed by a hierarchy of loops, but now they meet three at a point, Figure 28(d)—this is known as a Douady rabbit after the French mathematician, Adrien Douady. With c near $0.25 + 0.52i$ the loops meet four at a point and around $-0.5 + 0.55i$ they meet five at a point, Figure 28(e),(f). On the other hand, by choosing specific values of c such as $-i$ the loops collapse to give a *dendrite* or 'twig-like' Julia set, Figure 28(h).

As c becomes larger still, there is a further dramatic change in the nature of the Julia set—it ceases to be a connected entity and breaks up completely. The Julia set becomes a fractal dust, and is totally disconnected, so one cannot travel between any two distinct points of the set without going outside the set, for instance when $c = 0.66i$, Figure 28(g). In this situation, the Julia set is identical to the filled-in Julia set—any point not in the Julia set itself iterates to infinity, with no other destinations possible.

How can we make sense of what has been called the 'zoo' of Julia sets, which have many different characters that result from iterating the function $z \rightarrow z^2 + c$ for different complex numbers c? How can we tell which c will give a Julia set that is a single loop, with loops meeting 2, 3, or 4 at a time, or something else? The key to these questions is an even more complicated object, namely the Mandelbrot set.

The Mandelbrot set

To find the filled-in Julia sets, for each *fixed* complex number c we identified and coloured black the *various* initial points from which the itineraries under $z \rightarrow z^2 + c$ did not wander off to infinity. As an

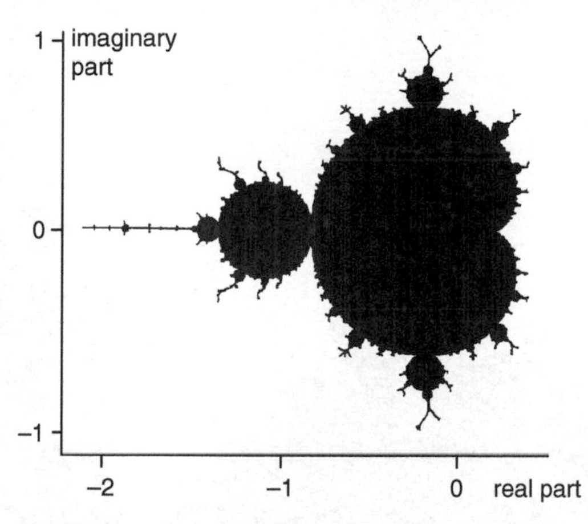

29. The Mandelbrot set comprises those complex numbers $a + bi$ in the blackened portion of the complex plane

alternative we could always take the *same* initial point and examine how the itineraries under $z \to z^2 + c$ change as *c varies*.

So, for each complex number $c = a + bi$ we now examine the itinerary under

$$z \to z^2 + c$$

starting at the origin $0 = 0 + 0i$. If the itinerary does not wander off to infinity we colour black the point $c = a + bi$ in the complex plane, in other words the point with coordinates (a,b). The set obtained in this way is shown in Figure 29—the *Mandelbrot set.* A rough picture may be computed by examining the iterates of 0 under $z \to z^2 + c$ for complex numbers c ranging across the plane—if the first 100 iterates remain close to the origin then c can be deemed to be in the Mandelbrot set.

Although defined by iterating functions of the simple form $z \to z^2 + c$, the Mandelbrot set is extraordinarily complicated, far more so than

30. Successive zooms on the Mandelbrot set

any individual Julia set. There is a prominent *cardioid* or 'heart-shaped region' with many nearly circular *buds* touching its perimeter. The largest bud is on the left of the cardioid and is exactly circular, the next largest buds are a pair located at the top and bottom of the cardioid, and there are many smaller buds around the cardioid. These in turn have much smaller buds surrounding them, and so on. But there is far more than just the buds. Very fine 'hairs' grow out from various points on the buds and within these hairs are minute copies of the entire Mandelbrot set. Close examination near the edge of the Mandelbrot set reveals a wealth of fine structure: high magnification shows shapes such as seahorses, multi-armed spirals, and windmills, Figure 30. In principle, one can continue to zoom into the boundary of the Mandelbrot set indefinitely and the variety of shapes to be found is unending.

How does this help understand the structure of the Julia sets? The following very simple statement, sometimes called the 'Fundamental Theorem of the Mandelbrot Set', gives the answer. It was proved, though not quite in these terms, independently by Gaston Julia and Pierre Fatou around 1919.

> The point c is in the Mandelbrot set precisely when the Julia set of the function $z \to z^2 + c$ is connected.

The word 'connected' means that the set is in a single piece, so that it is possible to trace a route between any two points of the set without leaving the set. If c lies outside the Mandelbrot set then the corresponding Julia set is totally disconnected or dust-like. The Mandelbrot set tells us about the connectedness of the Julia sets. There is a strong dichotomy, a Julia set is either connected or totally disconnected, the situation of several separate connected parts never occurs. Looking at the examples in Figure 28, all but one correspond to c inside the Mandelbrot set, with the remaining one, example (g), a totally disconnected dust.

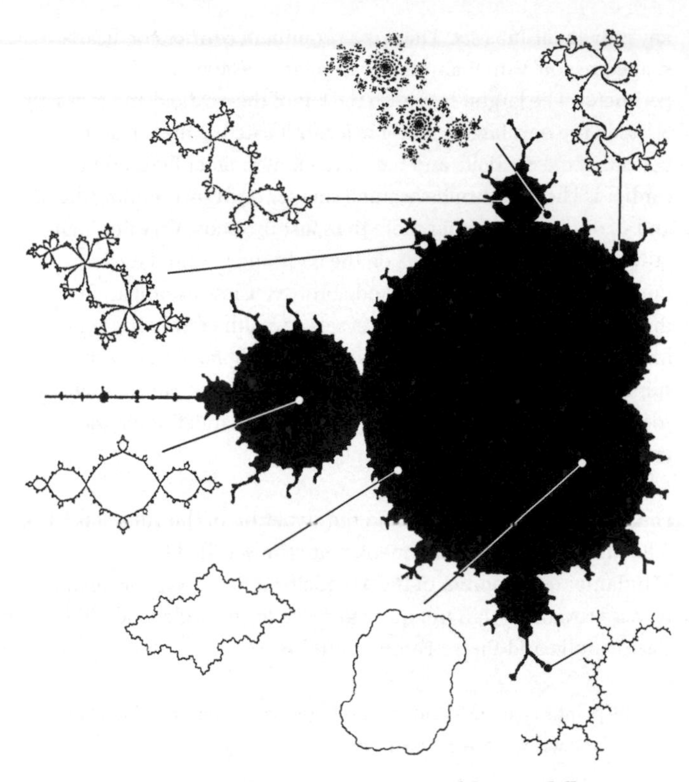

31. Julia sets corresponding to various points of the Mandelbrot set

This Fundamental Theorem is a remarkable statement. The Mandelbrot set was defined purely in terms of iteration. The Theorem tells us that the Mandelbrot set contains a great deal of *geometric* information—it codifies which of the vast range of Julia sets are connected. But this is just the start. Much more can be inferred from the Mandelbrot set about the structure of the Julia sets—it provides a directory of the various geometric forms. Figure 31 shows a selection of Julia sets for various complex numbers c with their position relative to the Mandelbrot set indicated. The form of the Julia set depends on the particular component of the Mandelbrot set in which c lies. For c within the main cardioid, the Julia set will always be a loop or closed circuit. As c crosses from the cardioid to the circular bud on its left side a 'bifurcation' occurs and the structure of the Julia set changes fundamentally to having infinitely many loops meeting in pairs. For all c within this bud the Julia sets have this same 'topological' form, that is they are just distorted versions of each other, always with the loops meeting in pairs. Similarly, if c lies in one of the large buds at the top or bottom of the cardioid we get a rabbit—a Julia set with loops meeting 3 at a point. The smaller buds surrounding the cardioid correspond to Julia sets with loops meeting 4, 5, 6, ... at each point.

Pictures of Julia sets for complex numbers c lying on, or very close to, the edge of the Mandelbrot set require very delicate computation. Some have a rather exotic structure, such as in Figure 32. Julia sets for c on the boundary of the Mandelbrot set are at the transition between connected and totally disconnected sets and can be highly intricate. Points for which the iterates of 0 under $z \to z^2 + c$ are 'eventually periodic', called *Misiurewicz points* after the Polish mathematician Michal Misiurewicz, must always lie on the boundary of the Mandelbrot set. For example, taking $c = -\mathrm{i}$ gives the itinerary

$$0 \to -\mathrm{i} \to -1-\mathrm{i} \to \mathrm{i} \to -1-\mathrm{i} \to \mathrm{i} \to -1-\mathrm{i} \to \mathrm{i} \to \ldots;$$

after two steps it just flips back and forth between $-1-\mathrm{i}$ and i, and the Julia set is a dendrite, Figure 28(h). For Misiurewicz points c where the period of the eventual repetition in the itinerary is large, the Julia set can be extremely intricate. Moreover, under very high magnification around such points, the boundary of the Mandelbrot set looks almost exactly like the Julia set itself—that is the Mandelbrot set boundary contains tiny pictures of the infinite gallery of exotic Julia sets! Since Misiurewicz points may be found arbitrarily close to every point on the boundary of the Mandelbrot set, its complexity becomes mind-boggling.

32. An exotic Julia set with $c = -0.6772 + 0.3245i$, a point very close to the boundary of the Mandelbrot set

Of course, with such irregularity, and the prevalence of small nearly similar copies of the Mandelbrot set within the hairs, it is natural to think of the Mandelbrot set as a fractal. This is not strictly true, since its area is positive (about 1.507 square units) and most of the set has no fine structure—zooming in on a point properly inside the cardioid or one of the buds just reveals a solid black area. However, the *boundary* of the Mandelbrot set is universally regarded as a fractal—it certainly has fine structure and highly complex quasi-self-similar features. But it was not until 1998 that a Japanese researcher Mitsuhiro Shishikura showed that, although the boundary has area 0, its dimension is 2, indicative of the extreme complexity of the set. The Mandelbrot set still harbours many mysteries. It is connected, but it is not known whether it is *locally* connected, roughly, whether it is always possible to travel between nearby points by short paths within the set, or whether there are pairs of points which require a disproportionately long route to get from one to the other.

The Mandelbrot set is defined in terms of a specific family of functions (3), but one might consider the form of Julia sets of other families of functions depending (in some reasonable way) on a complex parameter c. Then the set complex numbers c where the character of the Julia set of the function changes abruptly, for example from connected to totally disconnected, is known as the *bifurcation set* of the family. For many such families of functions, copies of the boundary of the Mandelbrot set are to be found throughout the bifurcation set. In this way the Mandelbrot set is *universal*, it crops up naturally when iterating other functions, not just for the family (3) discussed here.

Back to Julia sets

We have seen that the basic form of the Julia sets of $z \to z^2 + c$ depends on the location of c relative to the Mandelbrot set. As c moves around within the main cardioid or one of the buds of the Mandelbrot set, the Julia set distorts somewhat, but its overall form does not change. But more than this is true. The itineraries of $z \to z^2 + c$ behave in essentially the same way wherever c is within one of these regions. As we have noted, for c in the main cardioid, the Julia set is a loop or distorted circle. Under iteration, itineraries starting inside the loop move rapidly towards some fixed point independent of the starting point, whereas those starting outside

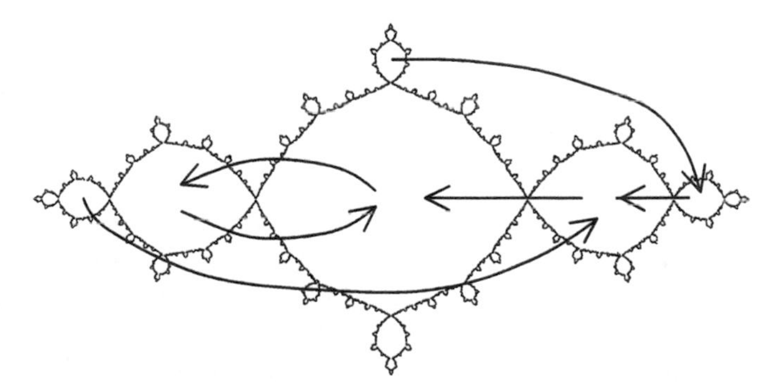

33. Progression between the loops of the Julia set for $z \to z^2 - 1$ under iteration

the loop shoot off to infinity. This description is valid for all c throughout the cardioid, regardless of the shape of the Julia set. So what if c lies in the large circular bud to the left of the cardioid? The Julia set consists of a main loop, surrounded by chains of smaller loops on either side, Figure 28(c). By definition of the Julia set, the itinerary starting at any point outside goes off to infinity. However, for an initial point inside one of the loops, each iteration takes a point from one loop to another. The progression is indicated by arrows in Figure 33. Once an itinerary reaches the largest loop in the centre of the Julia set it continues by moving back and forth between the main loop and the loop immediately to the left. Similarly, for c in other buds of the Mandelbrot set, the itineraries move between loops of the Julia set in a systematic manner.

Julia sets are fractals, so it is natural to ask about their dimensions. If $c = 0$ the Julia set is a circle, so has dimension 1. Elsewhere in the main cardioid, the loop becomes more wiggly and the dimension increases as c approaches the boundary of the Mandelbrot set. There are points on the boundary of the Mandelbrot set where the Julia sets have dimensions arbitrarily close to 2, corresponding to extremely complicated behaviour of the itineraries. However, once c leaves the Mandelbrot set the Julia set is a dust, which becomes very sparse and has small dimension when c is far away from the origin.

A historical note

Julia sets and the Mandelbrot set were studied long before their fractal form was appreciated. In 1915, the Académie des Sciences in Paris declared that the topic for its 1918 Grand Prix was to be the global properties of iteration of functions. The War was raging, and whilst serving with the French army, the mathematician Gaston Julia (1893–1978) received serious facial injuries. Nevertheless, he managed to continue his mathematical work in his hospital bed, and in December 1918 was awarded the

Prize. At the same time, another French mathematician Pierre Fatou (1878–1929) was working along similar lines, although he did not enter the competition. Thus in 1919, Julia and Fatou independently published major papers on the iteration of functions on the plane, including those of the form $z \to z^2 + c$. These papers contain the definitions of what are now known as Julia sets and the Mandelbrot set as well as including the Fundamental Theorem above. (Incidentally, the part of the plane not in the Julia set of a function is sometimes termed the Fatou set.) However, Julia and Fatou had little idea what these sets looked like, or even that they possessed any form of fine structure. There was no reason to suspect that the sets were particularly unusual; in any case with the calculating machines of the day it was not feasible to check the many iterates of the functions from many starting points that would be needed to observe fractality.

It was only when more powerful computers became available that reasonable images of these sets became possible. The first picture of the Mandelbrot set appeared in a paper published in 1978 by Robert Brooks and Peter Matelski. Their image was rather coarse, formed by lines of printed 'x's, with just 31 lines from top to bottom. The main cardioid and the three largest buds were apparent, along with the spike on the left hand side. It was in 1980 that Benoit Mandelbrot started to appreciate its extreme complexity. Mandelbrot worked at the IBM Thomas J. Watson Research Center in New York State and had enormous computing power (for the time) available. He produced much more detailed images of the set, with many more buds visible, and also computed enlargements of parts of the boundary to reveal the hairs, the mini-Mandelbrot sets, spirals, and other intricate features. He also obtained far higher resolution images of Julia sets than hitherto with many examples pictured in his 1982 *Fractal Geometry of Nature.*

On realizing the visual complexity of these objects, there was renewed energy amongst mathematicians to explain their structure, and the study of iteration of complex functions, which had progressed very little for 60 years, attracted the attention of top mathematicians. Adrien Douady and John Hubbard were leaders of this renaissance, and it was they who established many properties of the Mandelbrot set and proposed its name. Jean-Christophe Yoccoz (in 1994) and Curt McMullen (in 1998) were both awarded Fields Medals for research in the area, the greatest honour a mathematician can receive.

Chapter 5

Random walks and Brownian motion

A 'random walk' or 'drunkard's walk' is a simple idea that has many mathematical consequences and applications, and in particular leads to fractal graphs and other fractal objects. A walker sets out at time 0 from a point, which wc takc to be the origin, on a straight road. Each second he takes a step of length 1 unit (yard or metre, say) either forwards or backwards, the direction being chosen at random with a 50 per cent chance of going in each direction, perhaps by tossing a coin or just by a drunken whim. The graph in Figure 34 indicates the typical progress of the walker, showing his position each second in steps relative to the starting point.

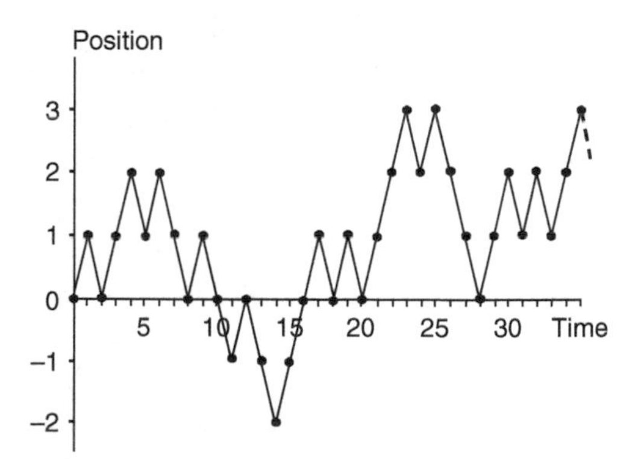

34. Progress of a typical random walk

With an equal chance of moving forwards or backwards at each step, the walker does not make very fast overall progress in either direction, but it is quite likely that after a time he will have travelled some way from the origin. Figure 35 shows the first 100 steps of several independent random walks—as time progresses, the range of locations of the various walkers spreads out on either side of the origin. The position after 20 steps of 130 independent random walkers is plotted in Figure 36. These have the 'bell-shaped' or 'normal' distribution which frequently occurs in statistical samples. Heads and tails being equally likely, most walkers will have taken roughly the same number of steps forwards as backwards and so will be unlikely to be more than about 5 steps from the starting point. However, one or two walkers will have thrown a run of heads or tails, and so will be rather further from the starting point. Since steps forward and backward are equally likely the *average* position of all the walkers will be at 0, but most walkers will be some distance away from the

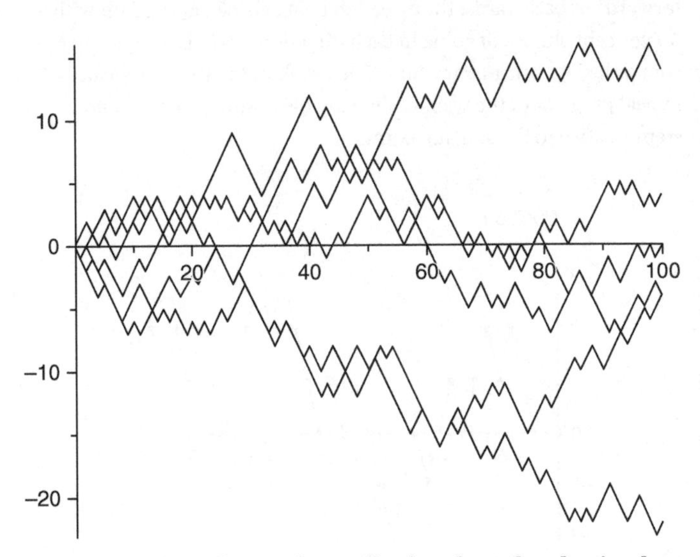

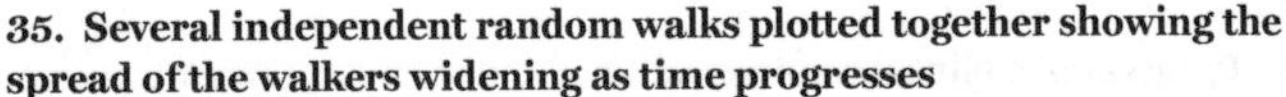
35. Several independent random walks plotted together showing the spread of the walkers widening as time progresses

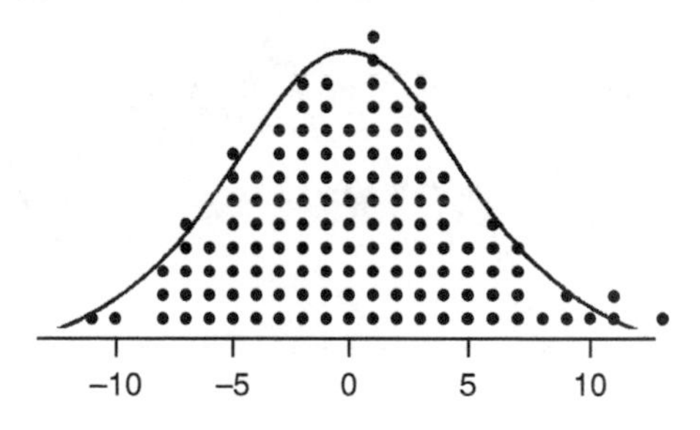

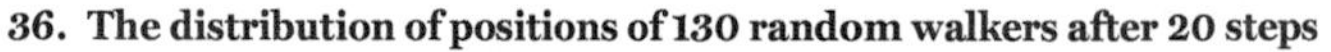

36. The distribution of positions of 130 random walkers after 20 steps

start, typically about 4–5 steps. To express this more precisely, the standard deviation, which gives the 'width' of the bell curve and indicates the 'typical' distance of a walker from the origin, is $\sqrt{20} = 4.47$. Continuing the random walk for a longer period of time, the average position after T steps will still be 0, but the 'typical' distance from the origin will increase and be about $\sqrt{T}$.

The graph in Figure 34 of the position of a random walker already has a rather spiky, irregular appearance. Now suppose that the walker takes rather shorter steps but much more frequently, say steps of length ½ every ¼ second or steps of length ¼ every $\frac{1}{16}$ second. The graph retains the same sort of overall appearance at a large scale, but smaller scale irregularity is also introduced (just as in the successive stages of the von Koch curve construction). By taking very small steps very rapidly the graph of the random walk takes on a fractal form, called the *Brownian* or *Wiener process*. For the process to remain interesting the steps taken at very small time intervals t need to be of length $\sqrt{t}$. (With shorter or longer steps the typical displacement from the origin, in other words the standard deviation, at any given time will become negligibly small or ridiculously large.) The Brownian process, Figure 37, has a fractal graph and, although random, always has box-counting dimension 1½. It is statistically self-similar, in the sense that enlarging a

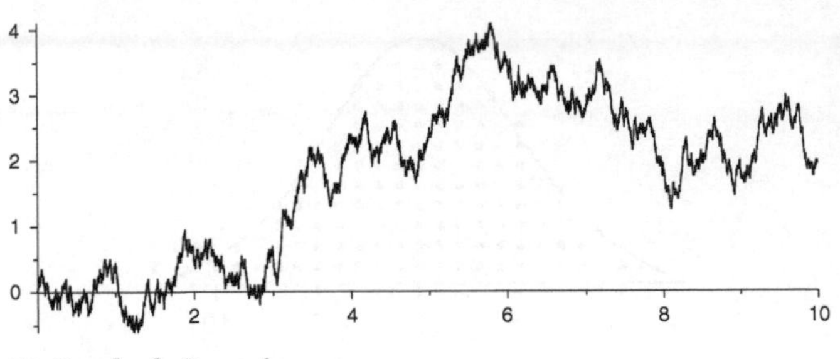

37. Graph of a Brownian process

small portion of the graph yields a graph of a similar overall appearance to the original, subject to statistical variations.

Many real phenomena which depend on very frequent random events appear very like a random walk or Brownian process. One striking example is of financial data such as share prices or exchange rates. As early as 1900, Louis Bachelier (1870–1946) proposed in his PhD thesis *Théorie de la spéculation* that share prices might behave in a similar way to a Brownian process. The prices of shares result from many individual investors estimating the future value of assets using information available to them at the time. Such information arrives randomly, and involves numerous factors, such as news or rumours of the behaviour of governments, companies, banks, etc. With virtually instantaneous trading possible, share prices are determined by a combination of a large number of small, seemingly random, upwards or downward steps in very short time intervals, with the consequence that a graph of share prices looks in many ways very similar to a Brownian graph. Indeed, graphs of share prices over a week, a year, or 20 years all have a very similar overall form, displaying strong statistical self-similarity. Moreover, box-counting on a graph of share prices invariably gives a dimension very close to $1\frac{1}{2}$, the dimension of the Brownian graph. There are mathematical and economic reasons for this. The stock market behaves in a way so as not to permit arbitrage, that is the possibility of an investor making a risk-free profit. Fractal models of share prices with graphs of dimensions other than 1½ may allow an arbitrage opportunity.

The Brownian process underlies option pricing in finance. A *call option* is a contract that gives the buyer the right, but not the obligation, to buy an agreed quantity of a stock or shares from the seller at a specified price, termed the *strike price*, on (or, in some cases, before) a specified expiry date. The buyer pays a certain amount to purchase the option, hoping that the stock will increase in value to above the strike price, so he or she can buy the stock for below its actual value on the expiry date and make an overall profit. If the buyer decides not to exercise the option and not buy the stock, then his or her loss is limited to the purchase price of the option. On the other hand, the seller does not expect such a rise in value, or perhaps is prepared to accept a reduced profit should the shares rise steeply in return for immediate cash. An obvious problem that arises is how to determine a fair price for the option so both buyer and seller feel that they have a good chance of making a profit, otherwise they are unlikely to agree an option contract.

A widely used basis for determining option prices is the Black-Scholes model (here a 'model' means a 'mathematical description') proposed in 1973 by Fisher Black (1938–95) and Myron Scholes (1941–) for which Scholes and Robert Merton (1944–), who developed the model, were awarded the Nobel Prize for Economics in 1997. The Black-Scholes differential equation describes how, under certain assumptions, the value of an option varies with time and with the price of the stock. A fundamental assumption is that the logarithm of the stock price follows a random Brownian process along with a fixed underlying drift (i.e. constant rate of increase or decrease). It is also assumed that there is no arbitrage opportunity, that is the model does not allow a risk-free profit, and that it is possible to borrow or lend cash at a known risk-free interest rate. The equation may be solved to obtain a formula for the 'natural' price for an option in terms of the time to expiry, the initial price of the stock, the strike price, and the risk-free rate. The natural price is determined by the requirement that the rate of return on the investment equals the risk-free rate, otherwise there would be an opportunity for arbitrage using the imbalance of the rates. A key idea in the derivation of the Black-Scholes equation is that terms relating to the underlying Brownian process cancel, so that there is no random component in the final equation and formula.

Despite its appeal, the Brownian description fails to reflect some crucial aspects of share behaviour—in particular it would essentially never permit the rapid changes that have occurred in reality, such as the stock market crashes of 1929, 1973, 1987, 2001, and 2007. A Brownian process requires that price changes over, say, each weekly period have a normal, 'bell-shaped' distribution with the typical weekly variation being given by its standard deviation. However, these market crashes would represent a fall of over 4 times the standard deviation, which with a normal distribution would occur less than every 300 years. Study of the actual weekly changes in prices shows that the variation is far from normally distributed and the Brownian, or random walk, model has nothing like the variation in step lengths needed to account for this. Consequently, many other mathematical processes that behave much more like real price data have been proposed, for example by allowing considerable variation in the step lengths of random walk, and even the possibility of instantaneous 'jumps' to occur from time to time. Another approach, that allows for the volatility of the markets to vary widely at different times, is to introduce a 'time warp', with the time scale on a Brownian, or other, graph distorted by stretching and compression in a highly irregular manner. Mandelbrot summed up the necessary features of a financial model in a nice way, by referring to the *Joseph effect* (from the biblical prediction of seven prosperous years followed by seven lean ones) and the *Noah effect* (the sudden dramatic effect of a flood). The Joseph effect reflects that trends tend to persist, but the Noah effect means that when things change they can change suddenly and dramatically. Plausible financial models need to allow for both effects.

A great deal of effort continues to be devoted to finding good mathematical descriptions of stock market behaviour, since a model that could indicate future trends might allow the discoverer to make large profits. Fractal models are natural candidates, being erratic in the short term but perhaps in the longer term having some underlying regularity that could be detected.

Random walks and Brownian motion in the plane or space

For a random walk in the plane, the walker starts at the origin and takes steps of a given length to the north, south, east, or west with equal probability, that is with a 25 per cent chance of moving in each direction. The path traversed will include many reversed steps, loops, and crossings, see Figure 38. (A practical setting for this might be the square grid of streets in New York with each 'step' comprising a walk between two adjacent junctions with a random choice of direction on reaching each junction.) The plane random walk has many surprising properties. For example, it can be shown that with probability one (i.e. with certainty), the path will *eventually* visit every point on the plane grid. However, the average time taken to reach any given point on the grid, or even to return to the start, is infinite!

Again, we can reduce the length of the steps and increase their frequency. Taking steps at very small time intervals t of length $\sqrt{t}$ in one of the four directions leads to a highly irregular continuous path in the plane that has many loops and self-crossings, Figure 39. This limiting path is called *planar Brownian motion* and is a statistically self-similar fractal of box dimension 2 but nonetheless occupying area 0, so is of negligible 'size' as a two-dimensional

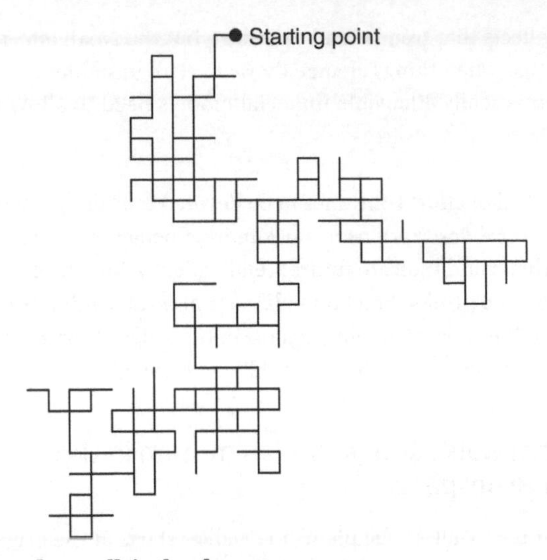

38. A random walk in the plane

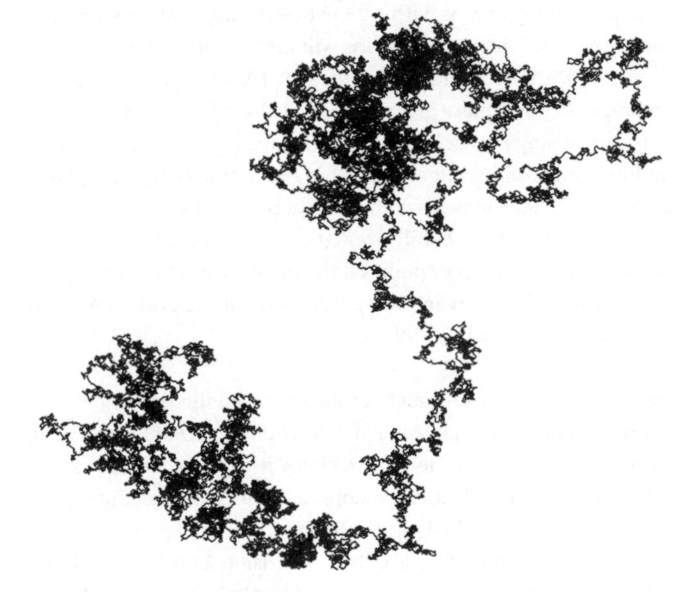

39. A Brownian path in the plane

object. Exactly the same procedure can be followed in three dimensional space, with each step going north, south, east, west, up, or down with equal likelihood to get a random walk in space. Taking very rapid steps of very short length gives *spatial Brownian motion*. This very irregular, tangled, path still has dimension only 2; in fact taking a photograph from any direction gives a picture that is precisely a realization of plane Brownian motion.

Brownian motion is named after the Scottish botanist Robert Brown (1773–1858) who in 1827 observed that minute particles of pollen in water moved on highly irregular paths. This was eventually explained as the result of the particles being continually bombarded in a random manner by molecules of water. Except at the finest scales, the paths followed in this 'real' Brownian motion are very similar indeed to those seen in the random walk construction. The effect of the particle being struck by molecules travelling in essentially random directions corresponds to the random choice of direction in the random walk. In 1905, Albert Einstein (1879–1955) showed that Brownian motion could be described by an equation of a similar form to the equation describing the diffusion of heat. This involved a term that could be related to measurable physical quantities and led to the determination of the size of atoms and the weight of molecules of various gases. Einstein's predictions were verified experimentally in 1913 by Jean Perrin (1870–1942) who won a Nobel Prize for this work. However, it was not until 1923 that Norbert Wiener (1894–1964) put Brownian motion on a completely rigorous mathematical basis.

Brownian motion is a very good example of a natural fractal phenomenon that can be explained very well by a simple mathematical model. Even so, the description inevitably breaks down at a very fine scale—a mathematical Brownian path continually changes direction to such an extent that it would require an infinite amount of energy for a particle of positive mass to follow such a path exactly.

Fractal fingering

Random walks and Brownian motion underlie a mechanism for fractal growth known as *fractal fingering* or, more formally, *diffusion limited aggregation*, which was proposed by physicists T. A. Witten and L. M. Sander in 1981. Fractal fingering occurs in a range of physical settings where there is some form of random diffusion, from forked lightning to bacterial growth on food.

Take a large disc in the plane, and site a small 'seed' in the centre. A small particle is released from a random point on the perimeter of the disc, and follows a random walk or Brownian path inside the disc until it touches the seed, whence it sticks to the seed. Another particle is released and moves randomly until it hits and sticks to the existing deposit. Further particles follow random paths and hit and stick to the existing cluster. Continuing with many thousands of small particles, the deposit grows outwards from the initial seed. Figure 40 indicates schematically how this process is simulated on a computer—the particles are squares on a very fine grid that follow a random walk on the grid, with equal probability of moving north, south, east, or west, until they stick at a grid square next to one that is already occupied. Although the growth is random, it assumes a characteristic appearance of repeatedly branching strands fingering outwards, Figure 41. These 'fractal fingers' or 'dendrites' tend to grow from near their ends, since when a new particle comes in on a random path it is much more likely to hit the cluster near its outer extremities than to weave its way in between the fingers to be deposited nearer the centre. The fine structure and fractal nature of such a growth is apparent, and box-counting estimates of the dimension are always close to 1.70, or about 2.43 for the analogous growth process in space.

Crystal growth by electrolysis is a good practical example of fractal fingering. A circular dish is filled with a shallow layer of

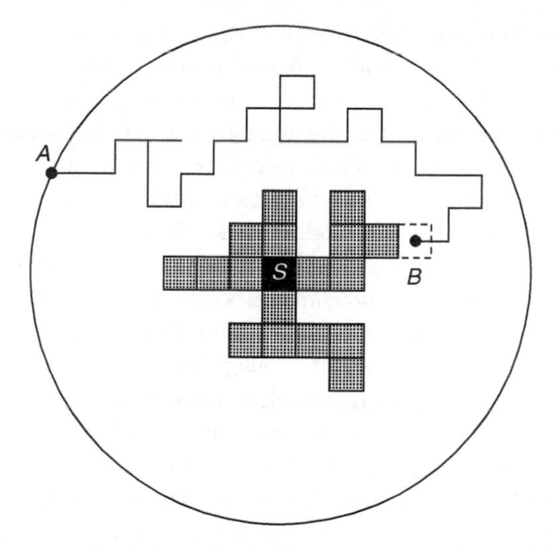

40. Computer simulation of fractal fingering: successive particles are released from random points *A* on the circle and perform random walks until they reach a square next to one already shaded, *B*. This square is shaded and the process repeated, so the shaded squares grow out from the initial 'seed' square *S*

41. A computer simulation of fractal fingering

copper sulphate ($CuSO_4$) solution, a copper wire is suspended in the centre of the dish, and a copper strip curved into a circle around the perimeter. If a battery of a few volts is connected with the negative to the centre wire and the positive to the perimeter wire, then a deposit of copper grows outwards from the centre and after an hour or so has a branched form very similar to Figure 41 perhaps with slightly thicker fingers. The chemical process taking place is very similar to that described above. In solution copper sulphate molecules split into copper Cu^{2+} ions and sulphate SO_4^{2-} ions which move around following random Brownian-like paths. When the voltage is applied the copper ions that hit the cathode receive two electrons and are deposited as copper. The copper deposit grows outwards with copper being deposited as the randomly moving copper ions hit the existing growth. Electrolysis of zinc sulphate ($ZnSO_4$) using a carbon electrode in the centre and zinc around the perimeter of the dish behaves in a like manner, with zinc fingers growing out from the centre.

Similar patterns can occur when bacteria grow on food. Nutrients, which encourage bacterial growth, diffuse across a suitable surface and are most likely to meet a bacterial colony near its extremities and so encourage growth in a branching, fingering manner. Under suitable conditions, the growth takes the form of fractal fingers. In a geological setting, fractal fingers of manganese oxide crystals sometimes form on rock surfaces when water rich in manganese flows through limestone or other rock and deposits tend to favour existing protrusions.

Lightning, which may be forked or fingered, is the massive electrical discharge that occurs when the electric field becomes strong enough to break down the normal insulating property of air. Branching electrical discharges with a fractal fingering form can also occur on the surface or interior of other insulating materials—these are known as Lichtenberg figures, named after the German physicist Georg Lichtenberg (1742–99) who observed them in 1777. They can be created experimentally by applying a high voltage source through a spark gap to a needle touching an insulating surface such as ebonite or glass. The discharge creates a fingering charge distribution across the surface which may be revealed by sprinkling flowers of sulphur and lead tetroxide onto the surface. A similar effect is obtained if a CD is put in a microwave cooker for three to four seconds (though this is not advised since it may damage the cooker!). Recently, spectacular fractal Lichtenberg figures have been created in blocks of clear acrylic by injecting a beam of electrons at high speed (over 95 per cent of the speed of light). This type of fractal fingering is also sometimes observed on the skin of people who have been struck by lightning, as a result of rupture of blood vessels under the skin as the lightning current spreads through the body.

Fractal viscous fingering may result from the interaction of two liquids under pressure. A Hele-Shaw cell (named after English engineer Henry Hele-Shaw (1854–1941)) consists of two glass plates placed about 0.5mm apart with the gap in between filled with a viscous liquid such as oil or glycerol. If water is injected through a small hole in one of the plates, then the water spreads through the oil in highly branched fingers as a result of the balance between the pressure and surface tensions of the liquids.

Although these settings may seem unconnected, they have much in common mathematically. Laplace's equation, named after the French mathematician Pierre-Simon Laplace (1749–1827), is a fundamental equation which describes many physical situations. The local density of copper ions in the copper sulphate solution, the pressure at points in the oil between the plates of a Hele-Shaw cell, and electrostatic potential all satisfy Laplace's equation away from the fingering itself, and, combined with the inherent randomness of the processes, this underlies the similarities between fractal growth patterns in different settings.

Fractal time records

As well as financial data, many other quantities vary with time in an erratic manner, for example, temperature and windspeed, discharge rates of rivers, levels of reservoirs, and internet traffic. Graphs of such quantities plotted over a long time range may display fractal behaviour, although the graphs may appear somewhat different from those of Brownian motion.

A method for analysing time records, known as *rescaled range analysis*, was introduced by the British hydrologist Harold Hurst (1880–1971) who made detailed observations over many years on the water levels of rivers and lakes, in particular on the Nile river system. His ultimate aim was to design reservoirs so that they neither overflowed nor dried up; this needed to take into account not only the normal variation in the water level, but also the extreme levels that might occur on rare occasions following extended periods of rain or drought. Hurst calculated the ratio over various time periods between the greatest variation in level and the typical variation, and noted that, to a good approximation, this ratio obeyed a power law in time. More precisely, for observations over a period of time t, let $R(t)$ denote the range (i.e. greatest minus the least values) and let $S(t)$ be the standard deviation, or typical variation, of the values. Hurst found that

$$\frac{R(t)}{S(t)} = \text{constant} \times t^H$$

for some power H between 0 and 1, where H is now termed the *Hurst index* or *Hurst exponent* of the observations. He also found that, as well as for water levels, this law held for many other classes of time-dependent data. Figure 42 shows graphs of simulated processes with Hurst indices of various values. For small H, the variation of the graph over short time intervals is almost as much as for much longer intervals, whereas for H near 1

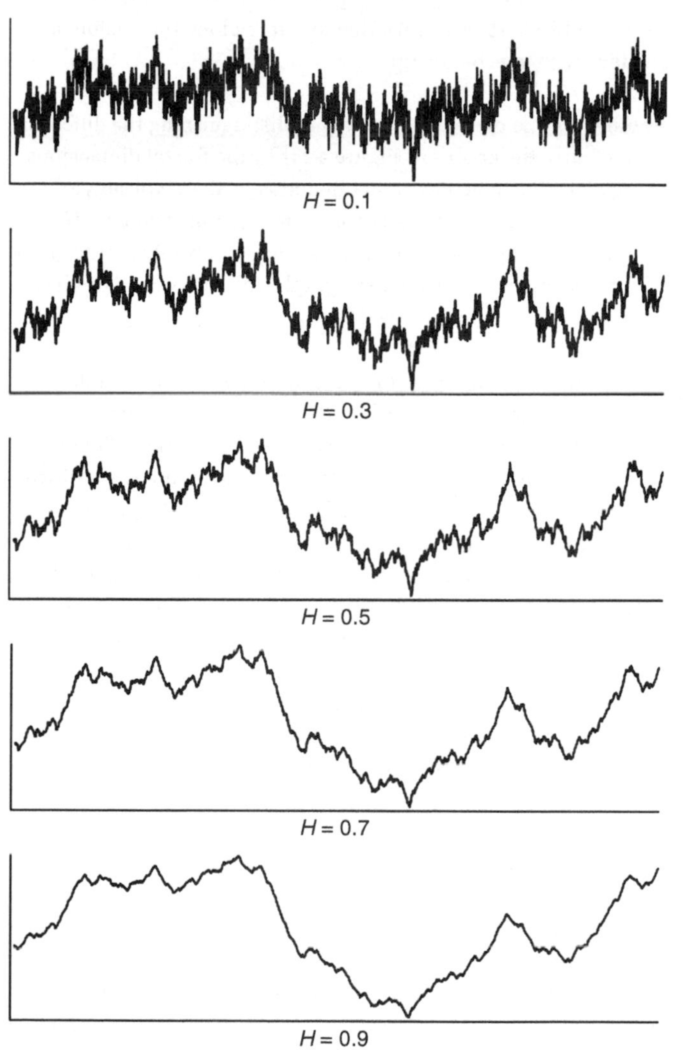

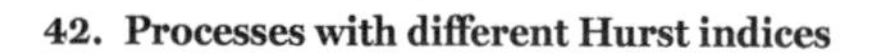
42. Processes with different Hurst indices

there is little short-term variation and it is a long time before a significant change builds up.

Another, maybe more obvious, way of distinguishing the different characters of the graphs of Figure 42 is by the fractal dimensions of the plots. It can be shown mathematically that, typically, a graph with Hurst index H has box-counting dimension $2 - H$. Thus a Hurst index close to 0 corresponds to a wildly erratic graph of dimension almost 2 whereas a graph with index close to 1 has dimension only slightly more than 1.

Perhaps the most important interpretation of the Hurst index is that it reflects the statistical predictability, or *long-term dependence* of phenomena that arise as a combination of many competing effects. If H is greater than 0.5 then there is a positive correlation between the past and future, that is if the process increases over a period then it is more likely to continue to increase for a while than to decrease. On the other hand, if H lies between 0 and 0.5 then following a period of increase a decrease is more likely. When $H = 0.5$, as in the case of the Brownian process, there is no historical memory—the future trend is independent of what has happened in the past. That share prices have Hurst index close to 0.5, or graphs of dimension 1.5, reflects the fact that past prices alone do not assist prediction of their future behaviour.

Hurst indices can provide a statistical indication of the likely extreme levels of a quantity over a period of time, but give little information as to when the extremes may occur. Nevertheless, this may be very useful: returning to Hurst's original hydrological context, such analysis can provide a basis for estimating the maximum level of a river likely over, say, the next 100 years and thus the height of flood defences likely to be needed.

In 1965, Hurst presented estimates of the Hurst index for a range of phenomena, including meteorological data, the thickness of varves (that is layers of mud deposited annually over a thousand or more years), sunspot numbers, and tree-ring widths, as well as hydrological data. The range of H for these natural phenomena was surprisingly narrow, taking values mainly between 0.68 and 0.77. Although recent work gives more widely ranging estimates, and indeed highlights the difficulty of estimating Hurst indices from empirical data, analyses do confirm values greater than 0.5, suggesting long-term dependence in many natural situations. This 'memory' can be explained in certain cases. For example, heavy rainfall not only has an immediate effect of filling streams and rivers, but it also saturates the drainage basin of a river, so the ground holds more water which will flow out into the river into the longer term and also becomes less able to store fresh rainfall.

Chapter 6
Fractals in the real world

Most of the fractals encountered so far in this book lie in the idealized world of the mathematician, where it is theoretically possible to repeat a construction step forever, or view an object at arbitrarily fine scales. Of course, the real world is not like that—in reality, we only encounter approximate fractals. If we zoom in too closely on a real object any self-similarity will be lost, and eventually we encounter molecular or atomic structure. Nevertheless, it can be very useful to regard natural objects as fractals if they exhibit irregularities or self-similarity when viewed over a significant range of scales. But all scientific descriptions or 'models' of reality are approximate, and this is a further instance. Planets are not perfect spheres but, for many purposes such as calculating their orbits, little accuracy is lost by assuming that they are. Similarly, Newtonian mechanics is completely adequate for most conventional calculations involving the structure or motion of everyday objects, though it breaks down at very small or very large scales, when quantum theory or relativity come into play.

Real objects tend to be described as 'fractal' if they contain parts that at two or more smaller scales appear in some way similar to the whole, indeed this is often enough for them to look very like mathematically precise fractals, such as for the fern computed in Figure 17. Alternatively, if estimates of dimension suggest a power law over a range of box-counts where the ratio of the largest to smallest sensible box sizes is more than about ten, this power law exponent may be stated as the dimension of the object and, provided that this is interpreted within the appropriate scale range, it can reflect useful information about the object and its physical behaviour.

Here is a selection of phenomena where the fractal viewpoint has provided valuable insight.

Coastlines and landscapes

Geographical shapes, such as coastlines, landscapes, and river networks display many fractal features. We saw in Chapter 3 that the box-counting dimension of the coast of Britain, estimated over a range of scales, is about 1.2 as opposed to the value of 1 that would be the case for a smooth boundary. It was Lewis Richardson (1881–1953) who first presented in a quantitative form the observation that the length of a coastline depends on the scale at which it is measured. Richardson was interested in the causes of warfare, and he collected a wide range of data that he thought might be relevant, including the lengths of certain coastlines and land frontiers that might affect tensions between neighbouring countries. He measured coastlines on maps using various step lengths to get estimates of the total length. Shorter step lengths detected more of the bays and capes that had been missed by coarser measurements, leading to a longer overall length. He realized that the calculated lengths increased considerably as the scale at which the measurements were made was reduced, and he displayed this on a graph relating the logarithm of the calculated length to that of the scale. The log-log graph gave a straight line but, whereas for a circle or smooth curve the line was flat, that is with slope or gradient 0, for the west coast of Britain the slope was about 0.25, providing a clear demonstration of the 'fractality' of the coastline. Richardson's statistics and graphs were not published until 1961, eight years after his death, although there is evidence that Richardson had talked of these effects many years before. This data was publicized by Mandelbrot in 1967 in a paper entitled *How Long is the Coast of Britain?* where he pointed out that Richardson's calculations essentially said that the coastline had dimension 1 + 0.25 = 1.25, valid over a wide range of scales. This article, which clearly demonstrated that fractional dimension was appropriate for describing a natural feature, played a key role in convincing scientists that such notions could be used to study real phenomena of an irregular nature.

Not surprisingly, fractals occur in other geographical features. The land surface of countries is an obvious example, which can display fractality from scales of hundreds of miles down to a few feet. Mountain ranges contain many peaks, each with subsidiary summits, with hillsides made up of large and small undulations resulting from local geology or erosion by streams, and smaller hillocks and tufts formed by soil irregularities. Of course, the dimension of the land surface varies from region to region, perhaps from about 2.25 for the Rockies to 2 (the dimension of a flat or smooth surface) for the Utah Salt Flats.

The form of a skyline is determined by that of the landscape, with higher hills and land obscuring lower features from view. A fractal landscape will present a fractal skyline; indeed the dimension of a skyline curve is typically exactly 1 less than that of the local land surface. Thus the dimension of a mountain skyline might be around 1.2 whereas for gentler countryside it would be closer to 1.

Mathematical fractal constructions have been used very effectively to simulate realistic looking landscapes. Methods used to generate random curves such as Brownian motion may be extended to produce random surfaces. One might start with a (low) pyramid and modify it by building on each face a smaller pyramid extending upwards or downwards a random amount. Repeating this procedure by replacing faces with ever smaller pyramids may give a fractal surface. Realistic landscapes may be produced by choosing the construction parameters appropriately, a surface of 'rural' appearance having dimension about 2.1 and a mountainous one having dimension up to 2.2. Fractal landscapes have been utilized widely in art and movies, with fractal planet and landscape simulations pioneered in the early 1980s films *Star Trek II: The Wrath of Khan* and *Star Wars: Return of the Jedi* and used in many later films. Usually it is the overall appearance of the landscape that is important rather than its precise form, and random fractal constructions can produce and vary such scenery very efficiently and at a much lower cost than creating elaborate film sets.

Turbulent fluids

A turbulent liquid or gas is one that behaves in a non-smooth, swirling fashion. Often when a tap is first turned on water emerges in a smooth stream, but then it breaks up into a gushing, irregular, turbulent, flow. Turbulent fluids can be difficult to control or predict and often have a violent association—a stream in spate, gusting winds, a bumpy aeroplane ride. Yet, despite having been studied intensively by scientists for hundreds of years, turbulence is far from understood. A first step in analysing turbulence would be to understand better the behaviour of solutions to the Navier-Stokes equations, the fundamental equations of fluid flow that were derived early in the 19th century. The importance and difficulty of this is indicated by the million dollar 'Millennium' prize offered by the Clay Mathematics Institute in the year 2000 for establishing some seemingly basic properties of the solutions of these equations.

One way turbulence may be realized is as a hierarchy or 'cascade' of eddies or whirls of decreasing sizes superimposed on each other. This idea, which was proposed by Lewis Richardson in 1922, encompasses a notion of self-similarity which is neatly summed up in his oft quoted parody:

Big whirls have little whirls that feed on their velocity,
And little whirls have lesser whirls and so on to viscosity.

(The original lines from Jonathan's Swift's *On Poetry: A Rhapsody* had previously been reworded, already with a fractal flavour, by the mathematician Augustus De Morgan (1806–71) as:

Great fleas have little fleas upon their backs to bite 'em,
And little fleas have lesser fleas, and so ad infinitum.)

In 1941 the Russian mathematician Andrey Kolmogorov (1903–87) formulated a quantitative theory based on this idea, applicable to relatively thin fluids. This depends on energy being passed down from the largest eddy via the intermediate ones, to reach the smallest eddies where the energy is dissipated as heat. Under the assumptions that the turbulence is isotropic, that is, there is no preferred direction, and homogeneous, that is, it fills the entire fluid, Kolmogorov developed a theory which fitted in reasonably well with physical observations. Nevertheless, there were observable differences, notably that when examined at a very fine scale the energy dissipation was not homogeneous, but varying considerably across the fluid in a rather clumpy manner, a phenomenon known as 'intermittency'. In the early 1970s, Benoit Mandelbrot proposed that this might be explained if the turbulent activity was not homogeneous, but was concentrated on a fractal region. Assuming such a model, experiments suggest that the turbulent region within the fluid should be a fractal of dimension around 2.4.

Fractals in our bodies

Some of the most extensive examples of branching fractal networks are to be found within the bodies of humans or other mammals, in particular in the breathing, blood circulation, and nervous systems. Whilst these remarkable structures lead to efficient physiological operation of the body, it is still not clear what evolutionary or biophysical mechanisms underlie this branching growth.

In the breathing or respiratory system, the windpipe, or trachea, splits into two bronchial tubes leading into the two lungs. These tubes split into narrower tubes, which continue to split repeatedly until, after about 11 levels of branching, they reach numerous very fine tubes called bronchioles which end in microscopic thin-walled sacs called alveoli. A lung contains around 400 million very closely spaced alveoli. Air breathed through the mouth or nose passes down the trachea and into the lung to reach the alveoli from whence oxygen is passed into the bloodstream and carbon dioxide is absorbed from the blood to be exhaled. Adult human lungs are about 12 inches long and 5 inches wide, but because of their branching fractal structure have an enormous surface area of about 100 square yards. It is the fractal structure that achieves this large area within a confined space and thus enables oxygen to be supplied to the blood efficiently and in adequate quantity for the whole body.

Blood vessels which transport blood through the body form another branching network. Blood is carried away from the heart by arteries leading to all parts of the body, and these divide into narrower arterioles and end in fine capillaries of about 0.01mm in diameter. Oxygen and nutrients pass out through the thin capillary walls to body tissue and waste products are absorbed into the capillaries. For this to function, every cell in the body must be within around 0.1mm of a blood vessel. Moreover, for the circulatory system to function efficiently, the distance between the heart and the capillaries should be no longer than necessary. This requires an intricate branching fractal network, comprising a total length of blood vessels of around 60,000 miles. In one component, the blood vessels in the retina of the eye have a striking branching structure, not unlike the fractal growth structure of Figure 41, and indeed the dimension of the network within the retina has been estimated at 1.7 over a fair range of scales.

Medical diagnosis

A graph of a person's heart rate measured every few seconds and plotted over a period of, say, half an hour, provides a great deal of information about the state of the heart. The heart does not beat at a constant rate: there is some variation in the rate in both the short term and longer term and the graph for a healthy heart has a fractal form not unlike the graph of share prices. This reflects the complex feedback system within the body, with information from all facets of its operation continually processed to determine the response from the heart rate. We have seen that even simple mathematical systems can lead to fractal outcomes, and the body is far from simple. Too erratic variation can indicate conditions such as atrial fibrillation when the chambers of the heart contract rapidly and randomly; on the other hand, too little variation can be a sign of congestive heart failure when the heart is unable to respond to the needs of the body. A fractal analysis of the heart beat graph can identify such problems. Very roughly, a healthy heart has a graph of dimension around 1.5 with lower or higher values indicating problems. Fractal analysis is used alongside other techniques, such as frequency analysis of heart signals, in diagnosing problems with the heart and blood circulation.

Considerable research has been done in the past few years to develop fractal methods for the early detection and classification of cancer tumours. Fractal analysis on scans of parts of the body and medical samples can indicate the development of tumour growth. For example, in vascular cancer, the arrangement of blood vessels, that is the vascular network, changes with the presence of tumoral cells which stimulate the production of the proteins needed to form new blood vessels. This can lead to the network becoming fragmented into a large number of very short branching links with a much more chaotic form than a healthy network. Fractal dimension can provide a good method of quantifying such changes, which are reflected in an increase in the box-counting dimension estimated over a suitable range of scales on images of the network. Fractal analysis has been incorporated into sophisticated automated methods of detecting a variety of cancers including breast and skin cancers and leukaemia.

Clouds

Clouds consist of microscopic droplets of water or crystals of ice suspended in the atmosphere. As warmer air rises from the earth it cools and eventually gets to a temperature at which it can no longer hold all its moisture. The surplus water vapour condenses as tiny drops on particles of dust or other material in the atmosphere to form visible clouds. The cooler air returns to earth, setting up convection or circulating air currents. Clouds appear in many characteristic shapes, which depend on the behaviour of the currents and the prevailing atmospheric pressure and temperature. Factors such as the proximity of mountains and nature of the ground covering further affect the highly complex climatic system.

Of the many types of cloud, several display fractal features. Cirrus clouds, consisting of ice crystals at high altitudes (above 17,000ft), have a stratified appearance, with groups of narrow feathery strands forming wider 'mare's tails'. Altocumulus clouds, formed by convection between 6,000ft and 20,000ft, may consist of many, quite closely packed, rounded heaps of greyish-white cloud, perhaps with blue sky showing in between, giving a 'mackerel sky' effect. There is a form of self-similarity with smaller, similarly shaped clouds nestling in between the larger ones.

Cumulus clouds, sometimes called 'fair weather clouds', appear as large, white, fluffy heaps, with irregular curvy sides and top, but with a fairly flat base usually below 6,000ft. Cumulus clouds have fractal boundaries, and box-counting on a photograph of a cumulus cloud gives the dimension of a cloud's outline as about 1.16. There are theoretical reasons to expect that the surface of a three-dimensional object will typically be one more than the dimension of its outline on a photograph. For example, a photograph of a sphere is bounded by a circle and a circle has dimension 1 corresponding to the spherical surface having dimension 2. In the same way we would expect the cloud surface to have dimension 1.16 + 1 = 2.16. By reflecting both the sun's rays and the heat coming up from the earth as well as absorbing heat and re-radiating it, clouds have a major effect on the temperature of the atmosphere and the earth's surface. As fractals, these clouds have a very large and convoluted surface, with far more exposed boundary available for heat absorption, radiation, and reflection than there would be if the cloud was, say, spherical. The fractality of clouds is a major factor in the immensely complex behaviour of the global climate.

Galaxies

The notion that the universe itself might have some form of hierarchical structure goes back at least 200 years, with astronomers such as John Herschel (1792–1871) and philosophers such as Immanuel Kant (1724–1804) suggesting that the portion of space that we can see around us might be replicated at much vaster scales. In 1907, Edmund Fournier d'Albe (1868–1933) published a remarkable book entitled *Two New Worlds* which proposed a hierarchical arrangement of stars and included a diagram of an idealized formation which was an exactly self-similar fractal dust. In the 1970s, with improved technology, it became possible to pinpoint many distant galaxies. Hubble's law states that the speed at which galaxies are receding from Earth is proportional to their distance away, and this speed can be measured fairly accurately from the redshift, that is the amount that light from the galaxies shifts towards the red end of the spectrum. This technique provides a picture of what the universe looks like millions of light years away (a light year is the distance that light travels in a year, some 6,000 million million miles).

The distribution of galaxies is fundamental to theories of cosmology, that is the evolution and structure of the universe.

By the early 1980s galaxies had been mapped up to distances of about 50 million light years away. Based on this data, Jim Peebles, Benoit Mandelbrot, and others proposed that galaxies might have a hierarchical, fractal distribution, rather than a uniform 'homogeneous' distribution across space. There is now general agreement amongst cosmologists that, over a range of 'relatively small' scales, maybe up to about 300 million light years, galaxies are distributed in a fractal-like fashion with a dimension in space of around 2. Thus billions of stars group together to make up galaxies, the galaxies form clusters, and the clusters are arranged in superclusters. However, views diverge about what the universe looks like at even larger scales, with some believing that fractality continues indefinitely but others claiming that there is a cut-off scale above which the universe becomes rather homogeneous with dimension close to 3. Very recent observations at scales of up to 3 billion light years seem to favour homogeneity, but fractal clustering still cannot be ruled out even at large scales.

Fractal antennae

Antennae, or aerials, collect and radiate radio waves, and are an essential part of radio and television transmitters and receivers and, more recently, mobile phones and GPS systems. Antennae design is a complex problem: antennae normally operate at a specific 'resonant' wavelength or frequency, and whilst they may be of some use at other wavelengths, their efficiency decreases markedly. Ideally, antennae have sizes that are comparable with the wavelength of operation, typically ½ or ¼ of the wavelength, but there may be constraints requiring a rather smaller size. For example, mobile phones typically operate at frequencies of 900–1,800 MHz, corresponding to wavelengths of about 17–33cm, but their antennae need to be much smaller than this.

To overcome these difficulties fractal antennae have been introduced. One approach is to use a fractal shape to fit a large length of material that can receive or transmit radio waves within a small space. We have seen that a good approximation to a von Koch or other fractal curve can be very long, so a conductor in the form of a fractal can achieve this easily. A second possibility is to make a conductor in the form of a self-similar fractal and get resonances at several wavelengths corresponding to the similarity ratios of the fractal. For example, the Sierpiński triangle comprises similar copies of itself at scales 1, ½, ¼, ⅛, 1/16,... and would have high efficiency at these multiples of some fundamental wavelength. This idea has led to the 'Sierpiński dipole', consisting of two Sierpiński triangles placed next to each other with the input at adjacent corners. Varying the construction parameters of the Sierpiński triangle can provide antennae suitable for other sequences of wavelengths. On the other hand, a randomly fingering fractal tree avoids any such exact scaling repetition, so may be appropriate as the basis for an antenna over a continuous range of wavelengths.

Multifractals

This account would be incomplete without at least some mention of multifractals—the idea that a widely varying distribution can lead to a whole hierarchy of fractals. For example, if a drop of coloured dye (food colouring will do) is put into water, then it spreads irregularly through the water—in some places the dye remains highly concentrated, elsewhere it is less intense and in some places almost negligible. The locations at which the dye has a given concentration density are far from regular, and maybe a fractal. Different density values will give different fractals whose dimensions might be found, and plotting a graph of these dimensions for various densities gives a 'spectrum', known as a *multifractal spectrum*, which codes information about the distribution of the dye though the water.

The following example, in which we analyse the decimal expansions of all the numbers between 0 and 1, illustrates the multifractal idea, with the distribution of decimal digits in numbers leading to a whole range of fractals. If we write down a 'typical' number, perhaps by choosing each digit at random, such as

0.40986721504290346318650372909536258741121853608794...

each of the digits 0,1,2,3,4,5,6,7,8,9 appears about 10 per cent of the time. For example, 9.994 per cent of the first 10 million digits of π are 0, and taking even more digits the proportion of 0s in a 'typical' number will become closer and closer to 10 per cent. Nevertheless there are plenty of numbers where this limiting proportion is very different. A number with 0s in alternate places like

0.601040507030801040203030907060 40...

has 0 occurring 50 per cent of the time, though such numbers are much less common than those with 10 per cent of the digits 0. So how 'large' are the sets of numbers between 0 and 1 with 10 per cent, 20 per cent, or 30 per cent, etc., of their digits 0s? It turns out that the set of numbers with any given proportion of 0s is a fractal, so the natural way to measure the size is by a fractal dimension. Each percentage defines a different fractal set of numbers and their dimensions may be calculated and are given in the following table (for technical reasons we need to use Hausdorff's definition of dimension here rather than box dimension).

Percentage of 0s	0	10	20	30	40	50	60	70	80	90	100
Dimension	0.95	1.0	0.98	0.93	0.86	0.78	0.67	0.55	0.41	0.24	0.0

Figure 43 displays a graph of these dimensions—the *multifractal spectrum* of these digit proportions. Note that the set of numbers with 10 per cent of their digits equal to 0 has dimension 1, such

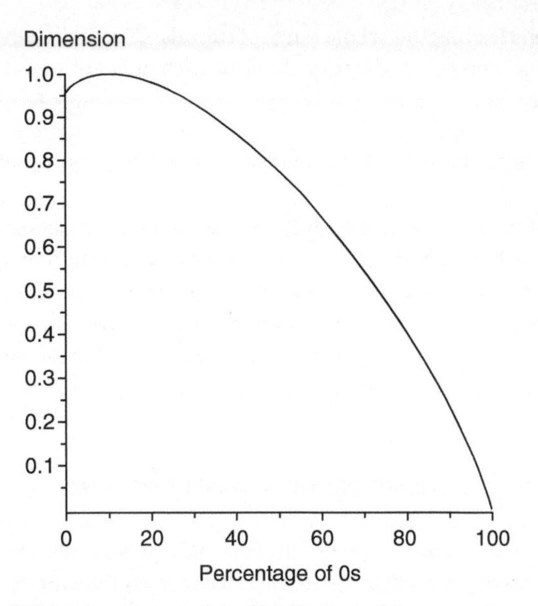

43. Multifractal spectrum of the distribution of the proportion of 0s in numbers

numbers being more ubiquitous than any other percentage of 0s. The central idea here is that a single quantity, in this case the proportion of 0s, is enough to define a whole range of fractals.

For another mathematical example, we might draw a fractal such as the Hénon attractor (see Figure 7) by plotting the iterates of a point under a function. Although the sequence of points fills up the fractal attractor, the density of points may vary considerably across the fractal, with some parts of the fractal visited by the itinerary far more often than others. Iteration gives a sequence of points distributed with varying density in the fractal rather than uniformly throughout the entire fractal. If we consider those parts of the attractor where the density of the iterates takes particular values, we get a hierarchy of fractals of varying dimensions that can be described by a multifractal spectrum.

One area where multifractal analysis has many applications is in image analysis. In a black and white photograph the shades range from solid black through many shades of grey to white, and this varying intensity defines a multifractal distribution. Multifractal image analysis is used to classify and distinguish textures of objects, to decompose images into regions of differing characters, and has even been used to identify different types of crops on satellite photographs. Automated multifractal analysis is also used on medical images, for example in the detection of cancerous tissues, and in particular is used in mammography for the diagnosis of breast cancer.

There are many other situations where widely varying distributions lead to multifractal spectra. Rainfall intensity may vary enormously across space and with time—under certain conditions there can be regions of drizzle, light rain, and heavy rain in close proximity giving a multifractal distribution. The density of distribution of a species of plant or tree across a country may have multifractal characteristics. Similarly, organisms such as phytoplankton (microscopic plant-like organisms in the ocean) can occur in concentrated clusters or be more dispersed, with their wildly varying density being multifractal.

Chapter 7
A little history

Geometry, with its highly visual and practical nature, is one of the oldest branches of mathematics. Its development through the ages has paralleled its increasingly sophisticated applications. Construction, crafts, and astronomy practised by ancient civilizations led to the need to record and analyse the shapes, sizes, and positions of objects. Notions of angles, areas, and volumes developed with the need for surveying and building. Two shapes were especially important: the straight line and the circle, which occurred naturally in many settings but also underlay the design of many artefacts. As well as fulfilling practical needs, philosophers were motivated by aesthetic aspects of geometry and sought simplicity in geometric structures and their applications. This reached its peak with the Greek School, notably with Plato (c 428–348 BC) and Euclid (c 325–265 BC), for whom constructions using a straight edge and compass, corresponding to line and circle, were the essence of geometric perfection.

As time progressed, ways were found to express and solve geometrical problems using algebra. A major advance was the introduction by René Descartes (1596–1650) of the Cartesian coordinate system which enabled shapes to be expressed concisely in terms of equations. This was a necessary precursor to the calculus, developed independently by Isaac Newton (1642–1727) and Gottfried Leibniz (1646–1714) in the late 17th century. The calculus provided a mathematical procedure for finding tangent lines that touched smooth curves as well as a method for computing areas and volumes of an enormous variety of geometrical objects. Alongside this, more sophisticated geometric figures were being observed in nature and explained mathematically. For example, using Tycho Brahe's observations, Johannes Kepler proposed that planets moved around ellipses, and this was substantiated as a mathematical consequence of Newton's laws of motion and gravitation.

The tools and methods were now available for tremendous advances in mathematics and the sciences. All manner of geometrical shapes could be analysed. Using the laws of motion together with the calculus, one could calculate the trajectories of projectiles, the motion of celestial bodies, and, using differential equations which developed from the calculus, more complex motions such as fluid flows. Although the calculus underlay all these applications, its foundations remained intuitive rather than rigorous until the 19th century when a number of leading mathematicians including Augustin Cauchy (1789–1857), Bernhard Riemann (1826–66), and Karl Weierstrass (1815–97) formalized the notions of continuity and limits. In particular, they developed a precise definition for a curve to be 'differentiable', that is for there to be a tangent line touching the curve at a point. Many mathematicians worked on the assumption that all curves worthy of attention were nice and smooth so had tangents at all their points, enabling application of the calculus and its many consequences. It was a surprise when, in 1872, Karl Weierstrass constructed a 'curve' that was so irregular that at no point at all was it possible to draw a tangent line. The Weierstrass graph might be regarded as the first formally defined fractal, and indeed it has been shown to have fractal dimension greater than 1.

In 1883, the German Georg Cantor (1845–1918) wrote a paper introducing the *middle-third Cantor set*, obtained by repeatedly removing the middle thirds of intervals (see Figure 44). The

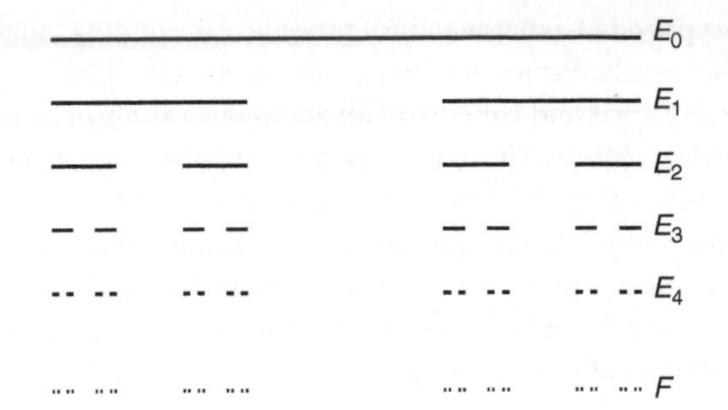

44. The middle-third Cantor set F. Each stage of the construction, $E_1, E_2, \ldots$ is obtained by removing the middle third of each interval in the preceding stage

Cantor set is perhaps the most basic self-similar fractal, made up of 2 scale ⅓ copies of itself, although of more immediate interest to Cantor were its topological and set theoretic properties, such as it being totally disconnected, rather than its geometry. (Several other mathematicians studied sets of a similar form around the same time, including the Oxford mathematician Henry Smith (1826–83) in an article in 1874.) In 1904, Helge von Koch introduced his curve, as a simpler construction than Weierstrass's example of a curve without any tangents. Then, in 1915, the Polish mathematician Wacław Sierpiński (1882–1969) introduced his triangle and, in 1916, the Sierpiński carpet. His main interest in the carpet was that it was a 'universal' set, in that it contains continuously deformed copies of all sets of 'topological dimension' 1. Although such objects have in recent years become the best-known fractals, at the time properties such as self-similarity were almost irrelevant, their main use being to provide specific examples or counter-examples in topology and calculus.

It was in 1918 that Felix Hausdorff proposed a natural way of 'measuring' the middle-third Cantor set and related sets, utilizing a general approach due to Constantin Carathéodory (1873–1950). Hausdorff showed that the middle-third Cantor set had dimension of log2/log3 = 0.631, and also found the dimensions of other self-similar sets. This was the first occurrence of an explicit notion of fractional dimension. Now termed 'Hausdorff dimension', his definition of dimension is the one most commonly used by mathematicians today. (Hausdorff, who did foundational work in several other areas of mathematics and philosophy, was a German Jew who tragically committed suicide in 1942 to avoid being sent to a concentration camp.) Box-dimension, which in many ways is rather simpler than Hausdorff dimension, appeared in a 1928 paper by Georges Bouligand (1889–1979), though the idea underlying an equivalent definition had been mentioned rather earlier by Hermann Minkowski (1864–1909), a Polish mathematician known especially for his work on relativity.

For many years, few mathematicians were very interested in fractional dimensions, with highly irregular sets continuing to be regarded as pathological curiosities. One notable exception was Abram Besicovitch (1891–1970), a Russian mathematician who held a professorship in Cambridge for many years. He, along with a few pupils, investigated the dimension of a range of fractals as well as investigating some of their geometric properties.

It was not until the mid-1960s that fractals started to become more widely studied in mathematics, science, and economics. One reason for this was the emergence of relatively powerful computers, enabling reasonable pictures of fractals to be produced, raising awareness of their beauty and intricacy. Benoit Mandelbrot (1924–2010) must take a great deal of credit for the development of many of the fractal ideas which are intensively studied and applied today. Mandelbrot was born in Warsaw, and, with the growing Nazi threat, fled to France with his family in 1936 where he studied and obtained a doctorate at the University of Paris in 1952. As well as working on mathematics he interested himself in areas of physics, astronomy, and economics with increasing excitement as he found self-similarity and its effects across a wide range of settings. In 1958, he joined the IBM

Research Division in Yorktown Heights, New York, where he had access to state of the art computers and where he remained until 1987 when he took up a position at Yale University.

By the late 1960s, Mandelbrot's thesis, that irregular objects should be regarded as the norm rather than the exception and deserved to be studied in a systematic and unified way, started to become better known and more widely accepted. Nevertheless, scepticism remained in some quarters of the mathematical and scientific community. Mandelbrot introduced the word 'fractal' in 1975. Derived from the Latin 'fractus' meaning 'broken', it was certainly a good choice—succinct and far less dry than 'sets of fractional dimension', the term hitherto used by mathematicians. Mandelbrot's 1975 book *Les objets fractals. Forme, hasard et dimension*, translated as *Fractals: Form, Chance and Dimension* in 1977, brought together notions of self-similarity from across science and mathematics and highlighted their widespread occurrence. Along with their 1982 successor, *The Fractal Geometry of Nature*, the books emphasized the need for fractal mathematics to be developed, and in some cases retrieved from almost forgotten papers such as those of Besicovitch, as well as applied in diverse settings. Mandelbrot highlighted the occurrence of self-similarity and fractality in a wide range of settings: galaxies, finance, topography, biology, chemistry, ... Further impetus came from pictures of exotic fractals, in particular the Mandelbrot set, that were starting to be produced and indeed regarded as an art form as well as complex mathematical objects.

Mandelbrot, who died in 2010, is often referred to as 'the Father of Fractals'. Since the 1980s virtually every area of science has been examined from a fractal viewpoint, and 'fractal geometry' has become a major area of mathematics, as a 'pure' subject of interest in its own right that continues to develop in parallel with its wide-ranging applications.

Appendix

This appendix sets out mathematical details of some properties that were taken on trust in the main part of the text and which may be of interest to readers a little more at home with algebra.

Powers and logarithms

Some of the dimension arguments in Chapter 3 involved raising numbers to powers other than whole numbers. To appreciate what this means, we must first consider whole number powers, obtained by repeatedly multiplying the number by itself. Raising a number to the power 2 is just squaring, so, for example, $5^2 = 5 \times 5 = 25$, where the superscript 2 indicates that 5 is raised to the power 2. To find the 3rd power, or cube, requires a further multiplication, so $5^3 = 5 \times 5 \times 5 = 125$. We can find any whole number power of a number by repeated multiplication, so in general $a^b = a \times a \times a \times \ldots \times a$ where a occurs b times in the multiplication, thus $5^6 = 5 \times 5 \times 5 \times 5 \times 5 \times 5 = 15{,}625$.

Grouping together terms in this product gives

$$5^6 = 5 \times 5 \times 5 \times 5 \times 5 \times 5 = (5 \times 5) \times (5 \times 5 \times 5 \times 5) = 5^2 \times 5^4$$

and similarly

$$5^6 = 5 \times 5 \times 5 \times 5 \times 5 \times 5 = (5 \times 5 \times 5) \times (5 \times 5 \times 5) = 5^3 \times 5^3$$

This illustrates the very useful *sum rule* for powers, that multiplying together two powers of the same number equals the number raised to the sum of the powers, in symbols

$$a^b \times a^c = a^{b+c}.$$

Now suppose we raise a number to one power and then raise the result to another power, for example, if we square 5 and then cube the answer,

$$(5^2)^3 = 5^2 \times 5^2 \times 5^2 = (5 \times 5) \times (5 \times 5) \times (5 \times 5) = 5 \times 5 \times 5 \times 5 \times 5 \times 5 = 5^6.$$

The new power 6 is the product of the powers 2 and 3. In the same way, for whole number powers in general, to raise a number to a power and then raise the result to another power we multiply the powers, that is

$$(a^b)^c = a^{b \times c},$$

termed the *product rule* for powers.

The key idea in defining powers of other numbers is to do so in such a manner that the sum and product rules remain valid, even when b and c are not whole numbers. So how must we define 5 raised to the power ½? For the product rule to hold we require $(5^{1/2})^2 = 5^{1/2 \times 2} = 5^1 = 5$. Thus $5^{1/2}$ is the number such that when squared gives 5, in other words the (positive) square root of 5, which is about 2.236. In the same way $5^{1/3} = 1.710$ and $5^{1/4} = 1.495$ are, respectively, the numbers which when cubed and raised to the fourth powers give 5, that is the cube and fourth roots of 5. In general, the powers ½, ⅓, ¼, . . . just represent square roots, cube roots, and fourth roots, etc., so that, when b is a positive whole number and a is positive, $a^{1/b}$ is the bth root of a, the number that when raised to the power b gives a.

Now let's think about fractional powers where the numerator is not 1—for example what does $5^{3/2}$ or $5^{1.5}$ mean? Noting that $3/2 = (1/2) \times 3$, if the product rule for powers is to hold we must have that

$$5^{3/2} = 5^{1/2 \times 3} = (5^{1/2})^3 .$$

But $5^{1/2}$ is just the square root of 5, and we can find its cube, so that

$$5^{1/2} = (5^{1/2})^3 = (2.236)^3 = 11.180.$$

In a similar way, for any fraction p/q and any positive number a we must define $a^{p/q}$ to be $(a^{1/q})^p$, thus the p/qth power of a number is its qth root raised to the power p.

Finally, not every positive number can be expressed as a fraction, for example pi, the ratio of the circumference to the diameter of a circle, is an 'irrational' number, that is, a never ending decimal $\pi = 3.14159265\ldots$. Nevertheless, we can find fractions as close as we like to π and, as above, raise a number to such fractional powers, to get arbitrarily close approximations to the πth power. For example, $5^{3.1416} = (5^{1/10000})^{31416} = 156.9944$ which is close to the actual value of $5^{\pi} = 156.9925\ldots$ which may be found by taking π to more decimal places. Defining powers of numbers in this way ensures that the sum and product rules of powers remain true, even if the powers are not whole numbers.

Logarithms, which were used in Chapter 3 and elsewhere, are very closely related to powers. Indeed, the product and power rules for logarithms are little more than the two rules for powers written in a different way. Recall that the *logarithm* of a number is the power to which 10 must be raised to give that number, thus $a = 10^c$ and $c = \log a$ are just different ways of expressing the same relationship.

To see why the product rule for logarithms holds, if $c = \log a$ and $d = \log b$, then by definition, $a = 10^c$ and $b = 10^d$. Multiplying, $a \times b = 10^c \times 10^d = 10^{c+d}$ by the sum rule for powers. Hence $c + d$ is the power to which 10 must be raised to give $a \times b$; in other words

$$\log(a \times b) = c + d = \log a + \log b.$$

For the power rule, if $c = \log a^b$ then, from the definition of logarithms, $a^b = 10^c$. Using the product rule for powers twice,

$$a = a^{b \times 1/b} = (a^b)^{1/b} = (10^c)^{1/b} = 10^{c \times 1/b} = 10^{c/b}.$$

From the definition of logarithms $\log a = c/b$, so multiplying by b gives

$$b \log a = c = \log a^b$$

which is the power rule for logarithms.

Squaring complex numbers

The claim in Chapter 4 that squaring a complex number squares its magnitude and doubles its angle needs some justification. To square a general complex number $z = x + y\mathrm{i}$ we multiply out the brackets, so that

$$\begin{aligned}(x + y\mathrm{i})^2 &= (x + y\mathrm{i}) \times (x + y\mathrm{i}) \\ &= (x \times x) + (x \times y\mathrm{i}) + (y\mathrm{i} \times x) + (y\mathrm{i} \times y\mathrm{i}) \\ &= x^2 + 2xy\mathrm{i} + y^2\mathrm{i}^2 \\ &= (x^2 - y^2) + 2xy\mathrm{i},\end{aligned}$$

using that $i^2 = -1$ so that $y^2i^2 = -y^2$. Thus, in coordinate notation, where the complex number $x + yi$ is thought of as the point in the plane with coordinates (x, y), the squaring function $z \to z^2$ is given by

$$(x, y) \to (x^2 - y^2, 2xy).$$

To find the magnitude of $(x + yi)^2$ we apply Pythagoras' Theorem to the coordinates, so the square of its magnitude is given by

$$\begin{aligned}(x^2 - y^2)^2 + (2xy)^2 &= (x^2)^2 - 2x^2y^2 + (y^2)^2 + 4x^2y^2 \\ &= (x^2)^2 + 2x^2y^2 + (y^2)^2 \\ &= (x^2 + y^2)^2\end{aligned}$$

which is just the square of the magnitude of $x + yi$. Thus squaring a complex number squares its magnitude.

To show that squaring a complex number doubles the angle is a little trickier. Whilst this can be deduced from standard formulae from trigonometry, here we give a direct algebraic approach. We first address the special case where $z = x + yi$ has magnitude 1, so that $x^2 + y^2 = 1$. Then the magnitude of z^2 is also 1. Consider the isosceles triangle with vertices 0, 1, and z, which have coordinates by (0, 0), (1, 0), and (x, y). The square of the distance between any pair of points in the plane is, using Pythagoras' Theorem, the sum of the squares of the differences of the individual coordinates. Thus the square of the distance from z to 1 is

$$\begin{aligned}(x - 1)^2 + (y - 0)^2 &= x^2 - 2x + 1 + y^2 \\ &= 2 - 2x,\end{aligned}$$

since $x^2 + y^2 = 1$. In the same way, consider the isosceles triangle with vertices 0, z, and z^2, or coordinates by (0, 0), (x, y) and $(x^2 - y^2, 2xy)$. The square of the distance from z^2 to z is

$$
\begin{aligned}
&(x^2 - y^2 - x)^2 + (2xy - y)^2 \\
&\quad = (x^2)^2 - 2x^2y^2 + (y^2)^2 + x^2 - 2x(x^2 - y^2) + 4x^2y^2 - 4xy^2 + y^2 \\
&\quad = (x^2)^2 + 2x^2y^2 + (y^2)^2 + x^2 + y^2 - 2x(x^2 - y^2) - 4xy^2 \\
&\quad = (x^2 + y^2)^2 + x^2 + y^2 - 2x(x^2 + y^2) \\
&\quad = 2 - 2x
\end{aligned}
$$

again using that $x^2 + y^2 = 1$. Thus these two isosceles triangles both have two sides of length 1 and the third side of length $\sqrt{2-2x}$ and so are congruent. This means that the two triangles have the same angles, and in particular the angle between the lines joining z and 1 to the origin equals the angle between the lines joining z^2 and z to the origin. Thus the angle of the complex number z^2 is double that of z in the case where z has magnitude 1.

Finally, observe that any complex number may be expressed as a real number times a complex number which has magnitude 1 and the same angle, for example $2 + 2i = 2\sqrt{2}\,(\tfrac{1}{\sqrt{2}} + \tfrac{1}{\sqrt{2}}i)$ where $\tfrac{1}{\sqrt{2}} + \tfrac{1}{\sqrt{2}}i$ has magnitude 1 and the same angle as $2 + 2i$. In this way we may write any complex number z as $z = rz_1$ where r is a real number and z_1 has magnitude 1. By the special case above, the angle of z_1^2 is double the angle of z_1. Since the angle of $z^2 = r^2z_1^2$ equals that of z_1^2 and the angle of $z = r\,z_1$ equals that of z_1 we conclude that the angle of z^2 is double the angle of z for all complex numbers.

Further reading

Several other books in this 'Very Short Introduction' series include material that complements or elaborates on topics mentioned in this book. Timothy Gowers' wide-ranging overview *Mathematics* includes a broad discussion of geometry and dimension. Peter Higgins' *Numbers* includes discussions on the real numbers that form the middle-third Cantor set and the complex numbers that underlie Julia sets. *Chaos* by Leonard Smith shows how fractals often arise from 'chaotic' systems.

Those wishing to study fractals further should certainly take a look at Benoit Mandelbrot's *The Fractal Geometry of Nature* (W. H. Freeman, 1982), which, apart from its historical significance, provides a broad scientific and philosophical overview of fractals, with little technical mathematics but with many illustrations that are remarkably fine for the early 1980s. There are several books covering fractals and chaos with mathematics kept to a basic level, including *Does God Play Dice?* (Penguin, 2nd edn, 1997) by Ian Stewart, author of many popular mathematics books, and *Fractals: Images of Chaos* by Hans Lauwerier (Penguin, 1991). *Introducing Fractals: A Graphic Guide* by Nigel Lesmoir-Gordon and Will Rood, illustrated by Ralph Edney (Icon Books, 2009), provides a short, entertaining overview of fractals with insightful cartoons on every page. *The Colours of Infinity*, edited by Nigel Lesmoir-Gordon (Springer, 2nd edn, 2010), contains very readable articles by a number of experts and comes with a stunningly illustrated DVD of the associated TV documentary (the DVD is also available separately).

There are plenty of texts for those wishing to venture more deeply into the mathematics of fractals, for example *Chaos and Fractals* by Heinz-Otto Peitgen, Hartmut Jürgens, and Dietmar Saupe (Springer, 2nd edn, 2004), *Chaos and Fractals: An Elementary Introduction* by David Feldman (Oxford University Press, 2012), *Fractals Everywhere* by Michael Barnsley (Dover, 3rd edn, 2012), and *Fractal Geometry: Mathematical Foundations and Applications* by Kenneth Falconer (John Wiley, 3rd edn, 2013).

Books on applications of fractals include Jens Feder's *Fractals: Physics of Solids and Liquids* (Springer, 1988). Fractals in finance are discussed in Benoit Mandelbrot's *The (Mis)Behaviour of Markets* (Profile Books, 2008). For those who want to experiment with computer pictures, *Exploring Fractals on the Macintosh* by Bernt Wahl (Addison Wesley, 1994), *The Science of Fractal Images*, edited by Heinz-Otto Peitgen and Dietmar Saupe (Springer, 1988), and *The Computational Beauty of Nature* by Garry Flake (MIT Press, 2000), present many approaches to creating fractals on a computer.

Many of these books include some historical background and *Classics on Fractals*, edited by Gerald Edgar (Westview Press, 2003) brings together translations of key mathematical papers, from Weierstrass to Hausdorff to Mandelbrot. A personal view of the development of fractals by Benoit Mandelbrot is contained in *The Fractalist: Memoir of a Scientific Maverick* (Pantheon Books, 2012), published two years after the author's death.

Websites

There are numerous websites, of varying quality and accuracy, relating to all facets of fractals. The Wikipedia article <http://www.en.wikipedia.org/wiki/Fractal> gives an overview which has many useful links, including one to a further Wikipedia page with a lengthy illustrated list of fractals by dimension. The course site at Yale University set up by Michael Frame, Benoit Mandelbrot, and Nial Neger <http://classes.yale.edu/fractals> covers a wide range of aspects and examples. For a wealth of information about the lives and work of the people mentioned in this book, as well as of many other mathematicians, see the *MacTutor History of Mathematics Archive* maintained at the University of St Andrews, <www-history.mcs.st-and.ac.uk>.

Index

Page numbers in **bold** type refer to definitions

D

E

F

G

H

I

J

K

L

M

N

O

P

ECONOMICS
A Very Short Introduction
Partha Dasgupta

Economics has the capacity to offer us deep insights into some of the most formidable problems of life, and offer solutions to them too. Combining a global approach with examples from everyday life, Partha Dasgupta describes the lives of two children who live very different lives in different parts of the world: in the Mid-West USA and in Ethiopia. He compares the obstacles facing them, and the processes that shape their lives, their families, and their futures. He shows how economics uncovers these processes, finds explanations for them, and how it forms policies and solutions.

> 'An excellent introduction . . . presents mathematical and statistical findings in straightforward prose.'
>
> Financial Times

www.oup.com/vsi

GALAXIES

A Very Short Introduction

John Gribbin

Galaxies are the building blocks of the Universe: standing like islands in space, each is made up of many hundreds of millions of stars in which the chemical elements are made, around which planets form, and where on at least one of those planets intelligent life has emerged. In this *Very Short Introduction*, renowned science writer John Gribbin describes the extraordinary things that astronomers are learning about galaxies, and explains how this can shed light on the origins and structure of the Universe.

www.oup.com/vsi

GAME THEORY

A Very Short Introduction

Ken Binmore

Games are played everywhere: from economics to evolutionary biology, and from social interactions to online auctions. Game theory is about how to play such games in a rational way, and how to maximize their outcomes. Game theory has seen spectacular successes in evolutionary biology and economics, and is beginning to revolutionize other disciplines from psychology to political science. This *Very Short Introduction* shows how game theory can be understood without mathematical equations, and reveals that everything from how to play poker optimally to the sex ratio among bees can be understood by anyone willing to think seriously about the problem.

www.oup.com/vsi

GENIUS

A Very Short Introduction

Andrew Robinson

Genius is highly individual and unique, of course, yet it shares a compelling, inevitable quality for professionals and the general public alike. Darwin's ideas are still required reading for every working biologist; they continue to generate fresh thinking and experiments around the world. So do Einstein's theories among physicists. Shakespeare's plays and Mozart's melodies and harmonies continue to move people in languages and cultures far removed from their native England and Austria. Contemporary 'geniuses' may come and go, but the idea of genius will not let go of us. Genius is the name we give to a quality of work that transcends fashion, celebrity, fame, and reputation: the opposite of a period piece. Somehow, genius abolishes both the time and the place of its origin.

www.oup.com/vsi

INFORMATION

A Very Short Introduction

Luciano Floridi

Luciano Floridi, a philosopher of information, cuts across many subjects, from a brief look at the mathematical roots of information - its definition and measurement in 'bits'- to its role in genetics (we are information), and its social meaning and value. He ends by considering the ethics of information, including issues of ownership, privacy, and accessibility; copyright and open source. For those unfamiliar with its precise meaning and wide applicability as a philosophical concept, 'information' may seem a bland or mundane topic. Those who have studied some science or philosophy or sociology will already be aware of its centrality and richness. But for all readers, whether from the humanities or sciences, Floridi gives a fascinating and inspirational introduction to this most fundamental of ideas.

'Splendidly pellucid.'

Steven Poole, The Guardian

www.oup.com/vsi

NOTHING
A Very Short Introduction
Frank Close

What is 'nothing'? What remains when you take all the matter away? Can empty space - a void - exist? This *Very Short Introduction* explores the science and history of the elusive void: from Aristotle's theories to black holes and quantum particles, and why the latest discoveries about the vacuum tell us extraordinary things about the cosmos. Frank Close tells the story of how scientists have explored the elusive void, and the rich discoveries that they have made there. He takes the reader on a lively and accessible history through ancient ideas and cultural superstitions to the frontiers of current research.

> 'An accessible and entertaining read for layperson and scientist alike.'
>
> **Physics World**

www.oup.com/vsi

THE LAWS OF THERMODYNAMICS

A Very Short Introduction

Peter Atkins

From the sudden expansion of a cloud of gas or the cooling of a hot metal, to the unfolding of a thought in our minds and even the course of life itself, everything is governed by the four Laws of Thermodynamics. These laws specify the nature of 'energy' and 'temperature', and are soon revealed to reach out and define the arrow of time itself: why things change and why death must come. In this *Very Short Introduction* Peter Atkins explains the basis and deeper implications of each law, highlighting their relevance in everyday examples. Using the minimum of mathematics, he introduces concepts such as entropy, free energy, and to the brink and beyond of the absolute zero temperature. These are not merely abstract ideas: they govern our lives.

> 'It takes not only a great writer but a great scientist with a lifetime's experience to explains such a notoriously tricky area with absolute economy and precision, not to mention humour.'
>
> **Books of the Year, Observer.**

www.oup.com/vsi

Human Rights

A Very Short Introduction

Andrew Clapham

An appeal to human rights in the face of injustice can be a heartfelt and morally justified demand for some, while for others it remains merely an empty slogan. Taking an international perspective and focusing on highly topical issues such as torture, arbitrary detention, privacy, health and discrimination, this *Very Short Introduction* will help readers to understand for themselves the controversies and complexities behind this vitally relevant issue. Looking at the philosophical justification for rights, the historical origins of human rights and how they are formed in law, Andrew Clapham explains what our human rights actually are, what they might be, and where the human rights movement is heading.